Springer Texts in Education

Springer Texts in Education delivers high-quality instructional content for graduates and advanced graduates in all areas of Education and Educational Research. The textbook series is comprised of self-contained books with a broad and comprehensive coverage that are suitable for class as well as for individual self-study. All texts are authored by established experts in their fields and offer a solid methodological background, accompanied by pedagogical materials to serve students such as practical examples, exercises, case studies etc. Textbooks published in the Springer Texts in Education series are addressed to graduate and advanced graduate students, but also to researchers as important resources for their education, knowledge and teaching. Please contact Yoka Janssen at Yoka.Janssen@springer.com or your regular editorial contact person for queries or to submit your book proposal.

Rola Khishfe

The Complexities of Socioscientific Issues in Education

Research-Based Insights and Practical Strategies

Springer

Rola Khishfe
American University of Beirut
Beirut, Lebanon

ISSN 2366-7672 ISSN 2366-7680 (electronic)
Springer Texts in Education
ISBN 978-3-031-88005-6 ISBN 978-3-031-88006-3 (eBook)
https://doi.org/10.1007/978-3-031-88006-3

This Springer imprint is published by the registered company Springer Nature Switzerland AG
The registered company address is: Gewerbestrasse 11, 6330 Cham, Switzerland

Acknowledgment

I am deeply grateful to everyone who played a role in bringing this book to life. To the educators, researchers, and experts who generously shared their knowledge—your insights have shaped and enriched these pages in ways I could not have imagined.

A special thank you to my students and colleagues, whose curiosity, thoughtful discussions, and fresh perspectives have continually challenged and inspired me. Your contributions have been essential in exploring how socioscientific issues can transform science education.

To my family and friends—your support, patience, and encouragement have been instrumental as I navigated the many stages of writing and revision.

I also acknowledge the broader scientific and educational community for your continued efforts in promoting critical thinking, ethical inquiry, and meaningful learning. This book reflects the collective work of those dedicated to preparing future generations for informed and responsible engagement with science and society.

Introduction

The modern science classroom (whether we are talking about school or higher education) has evolved into a dynamic environment in which students address complex and sometimes contentious socioscientific issues (SSI) that have a substantial impact on our world. Within the chapters presented, the complexities of these issues are explored. Each scenario serves as an overview of crucial concerns, providing students with a nuanced viewpoint as they navigate the complexities of science, ethics, and society. The scenarios seek to develop critical thinking abilities and enhance informed decision-making by encouraging students to think, evaluate, and support their viewpoints.

The first chapter explores the integration of SSI into science education to teach argumentation components and aspects of the nature of science (NOS). By incorporating SSI into scientific education, students acquire critical thinking skills, get a better knowledge of the link between science and society, and nurture fundamental competences for responsible citizenship. SSI also makes scientific knowledge more relevant and interesting for students, increasing motivation and providing real learning experiences. Furthermore, the chapter covers the significance of NOS, which involves the epistemology of science and the inherent values linked with scientific knowledge. Understanding NOS helps students appreciate the uncertain, empirical, and subjective aspects of scientific knowledge, resulting in a more nuanced view of science. In addition, the chapter underlines the importance of argumentation in scientific education, emphasizing how it fosters critical thinking, decision-making, and involvement with SSI.

Chapter 2 discusses the methodology and purpose. The process entails carefully researching and developing distinct SSI scenarios that include a wide range of scientific fields and contemporary ethical quandaries. Science professors validated these scenarios and adjusted for subject validity.

In the realm of biology, the third chapter delves into the study of living beings, namely biodiversity: the variety of species, habitats, and genetic variants on Earth. The importance of biodiversity preservation and the repercussions of its loss are discussed, with a focus on the fragile balance between human activities and the natural environment. As we explore the intricate web of life, it becomes evident that our actions have far-reaching consequences. It prompts us to reconsider the delicate equilibrium that sustains the diverse tapestry of life on our planet. Additionally, the agricultural sciences within this chapter are particularly interested in the reciprocal link between human beings and livestock production. The ethical, environmental, and nutritional components of food production decisions are examined, raising questions about our responsibility to ensure sustainable practices. With the origins of fish farming and the complexities of managing its ethical and sustainable components, the examination tackles global demand, environmental effect, and nutritional considerations. The chapter delves deeper into the interconnections between farming methods and biodiversity, with an emphasis on insect farming as an unconventional but potentially sustainable alternative. This brings forth a paradigm shift in our understanding of food production, challenging traditional methods and advocating for innovative approaches. Moreover, in the field of public health, interest is focused on continuing advancements with the COVID-19 vaccine. The complexities of vaccine research, distribution, and ethical concerns in this worldwide effort are revealed. As we grapple with the aftermath of a global pandemic, it underscores the interconnectedness of science, society, and ethics. Concurrently, the scientific and ethical aspects of in vitro fertilization are examined, with a focus on its transformational influence on reproductive technology and cultural norms. This not only broadens our understanding of medical advancements but also prompts reflection on the ethical implications of manipulating life at its earliest stages.

Moving to the fourth chapter, which examines physics, this section explores the fundamental laws that govern the universe. Solar panels are highlighted as a technological advancement that uses renewable energy to meet our growing energy needs. As we delve into the realm of sustainable energy, it becomes imperative to consider the broader implications of our energy choices. This exploration opens a dialogue on the intricate balance between technical breakthroughs, energy generation, and the ethical considerations that impact our actions.

In the discussion of nuclear energy, the examination scrutinizes its sustainability and contemplates alternative solutions. The complex interplay between technological innovation and ethical responsibility is evident in the evaluation of nuclear energy. This prompts us to question not only the feasibility of certain solutions but also the moral obligations associated with our pursuit of technological progress. The evolving landscape of energy sources challenges us to reassess our priorities and adopt a more holistic approach to sustainable development.

The fifth chapter delves into environmental science, using Lebanon as a case study to examine the intricate interplay between human actions and Earth's ecosystems. It highlights the critical relevance of environmental protection and the serious consequences of overfishing on aquatic ecosystems, prompting a deeper reflection on humanity's role as stewards of the planet. This chapter also explores water-related issues—such as shortages, desalination, wastewater reclamation, and pollution—emphasizing the profound relationship between water management and society's long-term sustainability. A focal point of the chapter is Lebanon's pressing waste management challenges, including the notorious garbage crisis and the dilemmas surrounding waste disposal methods. By situating these issues within Lebanon, the chapter illustrates the socioscientific dimensions of environmental conservation, touching on topics like deforestation, climate change, and the impacts of urbanization on ecosystems. These examples underscore the shared responsibility to address environmental degradation and the ethical obligations to mitigate its effects. Ultimately, this chapter calls for a collective reevaluation of lifestyles, consumption patterns, and resource utilization. It emphasizes the interconnectedness of environmental issues, urging readers to embrace sustainable practices that safeguard the delicate balance of our planet's ecosystems while addressing specific regional challenges.

The next chapter focuses on chemistry, where scientific development meets environmental concerns. The scenarios investigate how electric cars influence pollution levels, challenging standard notions of sustainable mobility. In the face of a changing climate, the exploration of sustainable transportation options is becoming increasingly vital. Furthermore, the scenarios offer an exploration of the pervasive problem of plastic pollution, delving into the environmental consequences of relying on synthetic materials. This examination prompts a critical examination of our reliance on plastics and calls for innovative solutions to address the environmental crisis. In addition, this chapter delves into the larger issue of air pollution, examining the chemical nature of pollutants and how they influence human health. The discussion on air pollution draws attention to the interconnected nature of environmental and human health. It emphasizes the need for interdisciplinary approaches that consider the scientific, ethical, and societal dimensions of pollution. As we grapple with the consequences of pollution, it becomes clear that addressing these challenges requires a comprehensive understanding of the complex relationships between chemical pollutants and their impact on both the environment and human well-being.

The seventh chapter brings together contemporary ethical dilemmas at the intersection of science, society, and human well-being. This chapter investigates the changing environment of education and employment, ranging from reinventing workplace relationships in the age of remote labor to reviewing teaching practices in public schools. The evolving landscape of education and employment presents ethical considerations that extend beyond the confines of the scientific realm. Moreover, ethical problems in scientific research take center stage, including sensitive issues like animal experimentation, genetic manipulation, chemical additives in crops, euthanasia, and the ethics of captive breeding and cloning. The examination of these ethical dilemmas prompts a critical evaluation of the ethical frameworks that guide scientific research and technological advancements. Additionally, environmental and social viewpoints on science are examined via the lens of rapid fashion, cannabis usage, and health and nutrition regulations such as imposing a 'fat tax' on junk food and the controversies surrounding artificial food coloring and additives. The examination of these diverse topics highlights the multifaceted nature of ethical considerations in contemporary society. It prompts us to question not only the scientific implications of these issues but also their broader societal impact. The intersection of science and social issues necessitates a nuanced understanding of the ethical dimensions that shape our collective future. As we navigate the complexities of rapid societal changes, it becomes evident that ethical considerations are integral to fostering a harmonious relationship between science, society, and human well-being.

The last chapter explores the significant implications and recommendations associated with integrating SSI into science education. Studies suggest a positive correlation between SSI and improved critical thinking skills, emphasizing the potential of SSI to deepen students' understanding of science and prepare them for informed citizenship. Recommendations include providing professional development for educators, developing inclusive instructional materials, advocating for SSI integration in curricula, and conducting ongoing research to assess effectiveness. These efforts aim to create more engaging science classrooms and empower students to actively participate in societal discourse.

Contents

Chapter 1
Socioscientific Issues as Tools to Teach Argumentation and Nature of Science

Abstract Socioscientific issues (SSI) play an important role in science education because they combine scientific concepts with ethical, social, and technical considerations. These concerns, such as genetic engineering and climate change, provide real-world challenges that need critical thinking, reasoning, and evidence-based decision-making. Incorporating SSI into science curricula promotes scientific literacy by engaging students in meaningful conversation, increasing their knowledge of the nature of science (NOS), and developing critical skills such as ethical reasoning and reflective thinking. The research shows a strong link between SSI, NOS, and argumentation, suggesting that specific education in these areas improves students' capacity to formulate arguments, counterarguments, and rebuttals. Studies show that SSI-based learning is helpful in improving students' grasp of scientific content while preparing them for active citizenship in an ever-changing environment. The use of SSI in science education closes the gap between theoretical knowledge and its societal applications, allowing students to critically analyze and negotiate challenging scientific and ethical quandaries. As scientific education takes a more holistic approach, SSI emerges as a critical foundation for promoting informed decision-making and life-long learning.

Keywords Socioscientific Issues (SSI) · Scientific literacy · Nature of Science (NOS) · Argumentation · Ethical reasoning · Science education · Critical thinking

1.1 Socioscientific Issues in Science Education

Socioscientific issues (SSI) are intricate, socially relevant, and contentious topics that encompass science, technology, and society (STS), involving not only the full scope of STS but also ethical elements, students' moral reasoning, and emotional development (Sadler et al., 2007; Zeidler et al., 2002). Examples such as genetic testing and global warming present real-world challenges with ethical implications (Sadler et al., 2007). SSI are typically characterized by ambiguity, conflicting values, and the need for decision-making, appearing in various contexts, including

R. Khishfe, *The Complexities of Socioscientific Issues in Education*, Springer Texts in Education, https://doi.org/10.1007/978-3-031-88006-3_1

environmental concerns, health challenges, and technological breakthroughs. The intentional use of scientific subjects in SSI encourages student engagement in conversations, deliberations, and debates. These topics are generally contentious, requiring moral thinking or the evaluation of ethical factors in decision-making (Zeidler & Nichols, 2009). The aim is for SSI to be personally meaningful, engaging, and involve evidence-based reasoning, providing a framework for understanding scientific material.

The incorporation of SSI into science education is instrumental in enhancing scientific literacy on multiple fronts. Firstly, SSI education serves as a catalyst for the development of critical thinking skills among students, empowering them to assess scientific information critically. As students grapple with the ethical dimensions of real-world issues, they not only engage with scientific content but also cultivate the ability to make informed decisions based on a comprehensive understanding of the societal implications of these issues. Secondly, teaching SSI enables students to comprehend the intricate relationship between STS. This holistic understanding is crucial for scientific literacy as it goes beyond rote memorization of facts, encouraging students to explore the interconnectedness of scientific advancements with broader societal contexts (Khishfe, 2023a). In essence, the inclusion of SSI in science education emerges as a crucial component in the development of scientific literacy, enabling students in to be informed citizens capable of positively contributing to the societal challenges they face.

To demonstrate the importance of integration of SSI into scientific education, scholars presented a variety of justifications. SSI can be employed to evaluate students' argumentation abilities, since it provides an ideal framework for expressing one's perspective, debating ethical issues, and making educated judgments (Al-Jrrah et al., 2013; Khishfe, 2012). In addition, SSI has been used to teach two pillars of scientific literacy (BouJaoude, 2002; Khishfe, 2014): the nature of science (NOS) (e.g., Khishfe, 2022) and argumentation (e.g., Khishfe, 2023a).

Moreover, SSI improves students' comprehension of scientific content, cultivates critical thinking skills and a sense of social responsibility (Zeidler et al., 2002), which prepares students for active participation in society (Sadler et al., 2007). SSI provides a rich context for science learning, making scientific knowledge more relevant to students' everyday lives and thereby promoting increased motivation and engagement in the subject matter (Kolstø, 2006).

Furthermore, the inclusion of SSI in science education contributes to the development of essential competencies such as ethical reasoning, decision-making, reflective thinking skills, discourse skills, and the ability to critically evaluate information, which are crucial for responsible citizenship (Al-Zunaidi & Al-Salim, 2020; Alsaaideh, 2016; El Arbid & Tairab, 2020; Grace & Ratcliffe, 2003). The integration of SSI aligns with the goals of science education by bridging the gap between abstract scientific concepts and their real-world applications, offering students a more meaningful and authentic learning experience (Grace & Ratcliffe, 2003).

Furthermore, several researchers (Almarashi & Alshahri, 2021; El Arbid & Tairab, 2020) recognize the rapid growth of scientific knowledge and its implications for societal challenges. These researchers emphasize the significance of incorporating SSI and STS into school science to address social challenges that interact with technological and scientific progress. At the same time, including such concerns will allow students to make connections between society and science (Almarashi & Alshahri, 2021). Environmental contamination, genetically engineered food, and cloning are all topics that can test students' scientific knowledge and ethical reasoning.

Lastly, contemporary science education should go beyond simply transmitting factual knowledge, emphasizing the importance of engaging students in discussions about complex scientific issues to prepare them for the challenges of an ever-changing world (Frhod & Abdulwahed, 2022). Teachers play an important role in instilling critical thinking and ethical concerns in pupils, and the use of SSI in such educational environments facilitates continued learning throughout students' lives. SSI training provides an excellent chance to broaden scientific knowledge, improve critical thinking, strengthen decision-making abilities, and stimulate scientific concept research (Al-Zunaidi & Al-Salim, 2020).

1.2 Nature of Science

Building upon the intricate relationship between SSI and the development of scientific literacy, the focus now shifts to a detailed exploration of NOS. NOS includes the epistemology of science, which is a method of comprehending and producing scientific knowledge (Pritchard, 2016), as well as the intrinsic values and beliefs linked with this knowledge and its evolution (Lederman, 2007). In contrast to ontology, which studies reality and the many elements and categories that comprise it (Slevitch, 2011), epistemology focuses on the processes of knowing. There are several approaches to conceptualizing or defining NOS, including the consensus view (Lederman, 2007), the family resemblance approach, which depicts NOS as a system (Irzik & Nola, 2014), the whole science approach (Allchin, 2011), and the features of science, which emphasize discussion and inquiry (Matthews, 2012). Despite the diversity of methods, researchers and educators agree on common characteristics of NOS that are important to K-12 kids and potentially teachable. These elements especially target views about the genesis and construction of scientific knowledge, making them epistemological rather than ontological (Chi, 2008). According to Lederman (2007), these aspects include: (1) scientific knowledge being empirically derived from observations, (2) scientific knowledge being tentative and subject to change, (3) scientific knowledge being subjective, taking into account the scientist's background and experience, (4) scientific knowledge being socially and culturally embedded, (5) scientific knowledge involving creativity and imagination, (6) distinctions between observations and inferences, and (7) distinctions between theories and laws. In the framework of SSI, three components of NOS—tentativeness, empiricism, and subjectivity—have been shown to be related to argumentation (Khishfe, 2023b), and hence were investigated in the current study. Tentative scientific knowledge means that it is not absolute and is subject to change based on available facts or reinterpretation of current data. Empirical scientific knowledge is produced from real-world observations, which are essentially dependent on evidence. Subjective scientific knowledge recognizes the impact of the scientist's background and experience on its development.

1.3 Argumentation

Now delving into the importance of argumentation in science education, Kuhn (2010) used the term "science as argument," describing science as a social activity in which argumentation plays an important role in communication and persuasion. In this view, scientific thought is intrinsically social, and debate helps to revitalize this social side of science. This concept is consistent with the current acceptance of argumentation as a critical emphasis in the scientific community, particularly because strengthening argumentation skills has become a prominent goal in science education (NRC, 2012).

An argument is defined as a claim supported by a justification (Erduran et al., 2004). Involving students in argumentation engages them in a cognitive process of comparison and contrast, allowing them to support assertions, produce counterarguments, and construct and reconstruct knowledge (Osborne, 2010). This research looks at three aspects of argumentation: argument, counterargument, and rebuttal. An argument is a claim that expresses one's perspective; a counterargument contradicts the first argument, and a rebuttal refutes the counterargument.

Argumentation is also viewed as a sort of informal reasoning in which learners draw conclusions from ambiguous principles, particularly when dealing with SSI (Teig & Scherer, 2016). It is seen as critical to decision-making, particularly in the field of SSI (Karpudewan & Roth, 2018). SSI, as ill-structured social science-based topics with diverse views, make a good venue for discussion. Students who participate in SSI can build arguments based on their daily life experiences and actively contribute to societal discourse on problems that are important to them (Sadler, 2009). Nussbaum et al. (2012) emphasized the dual character of arguments in different settings, noting that scientific contexts are more theoretical, whereas socioscientific situations are more practical. However, they emphasized the significance of adding a scientific dimension to socioscientific situations to improve scientific literacy.

Research indicates a link between argumentation and the NOS (e.g., Kutluca & Aydin, 2017; McDonald, 2010). The 2015 NARST international conference examined epistemology and classroom culture as problems for argumentation research (Henderson et al., 2018). This epistemological shift enables students to see scientific knowledge as a social endeavor that is subject to criticism and debate, challenging inflexible and absolute viewpoints.

1.4 The Use of SSI to Teach Argumentation and NOS Aspects

The utilization of SSI as an instructional approach for teaching argumentation and NOS aspects has been investigated in various studies.

In Khishfe's study (2023b), participants engaged with two sets of questions, addressing both argumentation and the NOS, within scenarios that address SSI. The study examined the influence of three NOS aspects (subjective, tentative, and empirical) as independent factors on argumentation components (argument, counterargument, and rebuttal) as dependent variables. The results demonstrated a positive relationship between arguing and NOS, revealing that the tentative, empirical, and subjective factors significantly predict argumentation components. Notably, variations were observed based on the specific SSI context, encompassing genetically engineered food or water fluoridation. These findings provided concrete evidence highlighting the intricate links between argumentation and NOS knowledge within the SSI framework.

Furthermore, in dealing with SSI, argumentation is acknowledged as a critical component of decision-making processes (Amgoud & Prade, 2009; Karpudewan & Roth, 2018). SSI are complex, socially rooted problems in the field of social research, characterized by diverse views and interpretations. Given the possibility of several perspectives on each given SSI, each supported by appropriate arguments, SSI emerges as an ideal platform for discussion. SSI is recognized as a basis for anchoring students' comprehension and practical applications within the framework of everyday life. It provides unique chances for students to participate in cognitive processes and mimic real-world discourse practices. As a result, students may base their arguments on real-life experiences, actively participating in social conversations about concerns and difficulties that are personal and relevant to their lives (Sadler, 2009).

Another study conducted by Khishfe (2023a) investigated the effects of using a linked learning technique for explicit teaching in NOS and explicit instruction in argumentation, as opposed to a non-connected approach, on students' knowledge of NOS and argumentation abilities. The study included 42 tenth-grade students who were randomly assigned into two groups: connected and non-connected. Both groups received instruction from the same teacher over a two-month period, with an emphasis on a plant-related lesson. The linked group got concurrent and interconnected explicit training on NOS and argumentation components, whereas the non-connected group received separate explicit instruction on NOS and argumentation. At the beginning and end of the study, an open-ended questionnaire and individual semi-structured interviews were used to assess participants' NOS comprehension and argumentation abilities in socioscientific situations. The results showed that the connected group outperformed the non-connected group in terms of argument generation and counter argumentation. Concerning NOS, both the connected and non-connected techniques showed an overall increase in participants' comprehension of NOS characteristics. Therefore, the significance of this work stems from its implications for integrating NOS and argumentation training into the normal science curriculum, particularly in the context of socioscientific concerns.

Additionally, Eastwood et al. (2012) conducted a study to look at the influence of two learning contexts, socioscientific issues-driven and content-driven, for explicit-reflective nature of science education on students' NOS concepts. Four groups of 11th and 12th-grade anatomy and physiology students participated, with two courses following a curricular sequence focused on SSI (SSI group) and two classes following a content-based sequence (Content group). Both groups were given an open-ended NOS questionnaire at the start and end of the school year, and their results were used to develop student profiles. Quantitative analyses were carried out to assess pre-instruction NOS concepts across groups, as well as pre- and post-intervention changes within and between groups. Both the SSI and Content groups showed considerable improvements in major NOS themes, with no significant differences in gains across the groups. However, a qualitative study of post-instruction replies found that students in the SSI group used examples to illustrate their views on the social/cultural dimensions of NOS. The findings confirm the usefulness of SSI settings in encouraging gains in students' NOS knowledge, and they show that these situations promote sophisticated concepts that merit additional investigation.

Moreover, Johnson et al. (2020) conducted a study and used the SSI framework to study how two educators viewed SSI situations while enrolled in a STEM education graduate program. Using a case study technique, a link was discovered between the specific SSI case and the level of scientific argumentation, as well as the instructors' teaching strategies. The data illustrate the successes that instructors experienced when integrating scientific argumentation using SSI examples into their planned instructional tactics. Notably, the engaged instructors displayed higher levels of claim making, reasoning, and challenging sources of information. This study of educators enrolled in a STEM education graduate program exemplifies the utilization of the Socioscientific Issues (SSI) framework as an instructional strategy for teaching argumentation. The engaged instructors in the study exhibited enhanced levels of claim making and reasoning, illustrating the effectiveness of incorporating SSI examples to foster robust scientific argumentation.

References

Al-Jrrah, Z. A., Khataibeh, A. M., & Bni Khalaf, M. H. (2013). Basic tenth grade students' arguments about socio-genetic issues in biology as related to reasoning patterns. *Jordan Journal of Educational Sciences, 9*(3), 307–318.

Allchin, D. (2011). Evaluating knowledge of the nature of (whole) science. *Science Education, 95*(3), 518–542.

Almarashi, A. K., & Alshahri, S. Z. G. (2021). Science, technology, society, and environment (STSE) issues included in the first intermediate grade science course in the Kingdom of Saudi Arabia. 94–67 ,(16)5 ,المجلة العربية للتربية النوعية. https://doi.org/10.33850/ejev.2021.138595

Alsaaideh, M. H. (2016). The socioscientific issues and sociology of science integrated in the science textbook of the eighth grade and students understanding level of it. *Studies in Educational Sciences, 43*(Supplement), 1967–1986.

Alsaaideh, M. H. A., Olimmat, A. B. R., & Alsaaideh, J. A. (2017). The effect of using the socio-scientific issues approach on the comprehension of current environmental problems in light of scientific interests among Al-Balqa Applied University Students. ,(41)*2* , مجلة كلية التربية- جامعة عين شمس, 193–125.

Al-Zunaidi, T. A. R., & Al-Salim, G. H. I. S. (2020). Effectiveness of a program for teaching socio-scientific issues based on an ethical analysis approach on developing reflective thinking skills among high school students in Riyadh. 24)*4* ,مجلة العلوم التربوية) (Part 3), 71–122.

Amgoud, L., & Prade, H. (2009). Using arguments for making and explaining decisions. *Artificial Intelligence, 173*(3–4), 413–436.

BouJaoude, S. (2002). Balance of scientific literacy themes in science curricula: The case of Lebanon. *International Journal of Science Education, 24*, 139–156.

Chi, M. T. H. (2008). Three types of conceptual change: Belief revision, mental model transformation, and categorical shift. In S. Vosniadou (Ed.), *Handbook of research on conceptual change* (pp. 61–82). Hillsdale, NJ: Erlbaum.

Eastwood, J. L., Sadler, T. D., Zeidler, D. L., Lewis, A., Amiri, L., & Applebaum, S. (2012). Contextualizing nature of science instruction in socioscientific issues. *International Journal of Science Education, 34*(15), 2289–2315.

El Arbid, S. S., & Tairab, H. H. (2020). Science teachers' views about inclusion of socio-scientific issues in UAE science curriculum and teaching. *International Journal of Instruction, 13*(2), 733–748. https://doi.org/10.29333/iji.2020.13250a

Erduran, S., Simon, S., & Osborne, J. (2004). TAPping into argumentation: Developments in the application of Toulmin's argument pattern for studying science discourse. *Science Education, 88*(6), 915–933.

Frhod, W., & Abdulwahed, A. A. (2022). Decision making in socioscientific issues of biology teachers for secondary stage. اشراقات تنموية مجلة علمية محكمة *(Illuminations of Development, a Peer-Reviewed Scientific Journal), 34*, 203–228. Al-Qadisiyah University/College of Education.

Grace, M., & Ratcliffe, M. (2003). Making decisions about biological conservation issues in peer group discussion. In D. Psillos, P. Kariotoglou, V. Tselfes, E. Hatzikraniotis, G. Fassoulopoulos, & M. Kallery (Eds.), *Science Education Research in the Knowledge-Based Society* (pp. 241–247). Springer. https://doi.org/10.1007/978-94-017-0165-5_26

Henderson, J. V., Squires, T., Storeygard, A., & Weil, D. (2018). The global distribution of economic activity: Nature, history, and the role of trade. *The Quarterly Journal of Economics, 133*(1), 357–406.

Irzik, G., & Nola, R. (2014). New directions for nature of science research. In M. R. Matthews (Ed.), *International Handbook of Research in History, Philosophy and Science Teaching* (pp. 999–1021). Dordrecht, The Netherlands: Springer.

Johnson, J., Macalalag, A. Z., & Dunphy, J. (2020). Incorporating socioscientific issues into a STEM education course: Exploring teacher use of argumentation in SSI and plans for classroom implementation. *Disciplinary and Interdisciplinary Science Education Research, 2*(1), 1–12. https://doi.org/10.1186/s43031-020-00026-3

Karpudewan, M., & Roth, W. M. (2018). Changes in primary students' informal reasoning during an environment-related curriculum on socio-scientific issues. *International Journal of Science and Mathematics Education, 16*, 401–419.

Khishfe, R. (2012). Relationship between nature of science understandings and argumentation skills: A role for counterargument and contextual factors. *Journal of Research in Science Teaching, 49*(4), 489–514.

Khishfe, R. (2014). A reconstructed vision of environmental science literacy: The case of Qatar. *International Journal of Science Education, 36*(18), 3067–3100. https://doi.org/10.1080/09500693.2014.951980

Khishfe, R. (2022). Nature of science and argumentation instruction in socioscientific and scientific contexts. *International Journal of Science Education, 44*(4), 647–673.
Khishfe, R. (2023a). Connected learning: An approach for teaching nature of science aspects and argumentation components. *Journal of Research in Science Teaching, 60*(10), 2180–2215. https://doi.org/10.1002/tea.21855
Khishfe, R. (2023b). Relationship between nature of science and argumentation: A follow-up study. *International Journal of Science and Mathematics Education, 21*(4), 1081–1102. https://doi.org/10.1007/s10763-022-10307-0
Kolstø, S. D. (2006). Patterns in students' argumentation confronted with a risk-focused socio-scientific issue. *International Journal of Science Education, 28*(14), 1689–1716.
Kuhn, D. (2010). Teaching and learning science as argument. *Science Education, 94*(5), 810–824.
Kutluca, A., & Aydin, A. (2017). Changes in pre-service science teachers' understandings after being involved in explicit nature of science and socioscientific argumentation processes. *Science and Education, 26*, 637–668.
Lederman, N. G. (2007). Nature of science: Past, present, and future. In S. K. Abell & N. G. Lederman (Eds.), *Handbook of research on science education* (pp. 831–879). Lawrence Erlbaum Associates.
Matthews, M. R. (2012). Changing the focus: From nature of science (NOS) to features of science (FOS). In M. S. Khine (Ed.), *Advances in nature of science research: Concepts and methodologies* (pp. 3–26). Dordrecht, The Netherlands: Springer.
McDonald, C. V. (2010). The influence of explicit nature of science and argumentation instruction on preservice primary teachers' views of nature of science. *Journal of Research on Science Teaching, 47*(9), 1137–1164.
National Research Council. (2012). A framework for K-12 science education: Practices, crosscutting concepts, and core ideas. Washington, DC: National Academies Press.
Nussbaum, E. M., Sinatra, G. M., & Owens, M. C. (2012). The two faces of scientific argumentation: Applications to global climate change. In M. Khine (Ed.), *Perspectives on scientific argumentation* (pp. 17–37). Springer.
Osborne, J. (2010). Arguing to learn in science: The role of collaborative, critical discourse. *Science, 328*(5977), 463–466.
Pritchard, D. (2016). Epistemology. Palgrave Macmillan, London. https://doi-org.ezproxy.aub.edu.lb/10.1007/978-1-137-52692-2
Sadler, T. D., Barab, S. A., & Scott, B. (2007). What do students gain by engaging in socio-scientific inquiry? *Research in Science Education, 37*, 371–391. https://doi.org/10.1007/s11165-006-9030-9
Sadler, T. D. (2009). Situated learning in science education: socio-scientific issues as contexts for practice. *Studies in Science Education, 45*(1), 1–42.
Slevitch, L. (2011). Qualitative and quantitative methodologies compared: ontological and epistemological perspectives. *Journal of Quality Assurance in Hospitality and Tourism, 12*(1), 73–81.
Teig, N., & Scherer, R. (2016). Bringing formal and informal reasoning together—A new era of assessment? *Frontiers in Psychology, 7*, 1097.
Zeidler, D. L., & Nichols, B. H. (2009). Socioscientific issues: Theory and practice. *Journal of Elementary Science Education, 21*(2), 49–58. https://doi.org/10.1007/BF03173684
Zeidler, D. L., Walker, K. A., Ackett, W. A., & Simmons, M. L. (2002). Tangled up in views: Beliefs in the nature of science and responses to socioscientific dilemmas. *Science Education, 86*, 343–367.

Chapter 2
Purpose and Methodology

Abstract This chapter describes the study's purpose and methodology, which investigates the use of SSI in science education to improve students' knowledge of the NOS and argumentation skills. Given the importance of SSI in promoting scientific literacy, the study focusses on including a variety of scenarios from biology, physics, environmental science, chemistry, and ethical dilemmas. The process entails doing a thorough search of scholarly resources, including Google Scholar and Shamaa, to identify relevant SSI scenarios that address real-world issues such as pollution and COVID-19 vaccination. These scenarios were divided into scientific fields and reviewed by six science instructors to guarantee clarity and applicability. By including students in discussions about these issues, the study aims to improve their critical thinking and decision-making skills while emphasizing the interconnectedness of science and society.

Keywords Nature of science (NOS) · Socioscientific issues (SSI) · Science education · Critical thinking · Inquiry-based learning

2.1 Purpose

Socioscientific Issues (SSI) can serve as a powerful tool for teaching argumentation and nurturing an understanding of NOS. Including SSI in the classroom encourages students to explore the connection of advancements in science to the broader social contexts, which improves their scientific literacy (Khishfe, 2023a). Studies in various Middle Eastern countries, including Jordan, Lebanon, United Arab Emirates (UAE), Kingdom of Saudi Arabia, and Iraq, emphasize the significance of incorporating SSI into science curricula to address societal challenges, cultivate decision-making abilities, and prepare students for the demands of the twenty-first century (Almarashi & Alshahri, 2021; Alsaaideh, 2016; Al-Zunaidi & Al-Salim, 2020; BouJaoude, 2002; El Arbid & Tairab, 2020; Frhod & Abdulwahed, 2022; Khishfe, 2012, 2014, 2022, 2023a).

R. Khishfe, *The Complexities of Socioscientific Issues in Education*, Springer Texts in Education, https://doi.org/10.1007/978-3-031-88006-3_2

Several studies, such as Khishfe's (2023b) investigation of a linked learning approach and Eastwood et al.'s (2012) examination of learning contexts, highlight the positive impact of incorporating SSI towards the instruction of NOS and argumentation. The study by Khishfe (2023b) demonstrates a positive relationship between arguing and NOS understanding, showing that the NOS aspects of tentativeness, empiricism, and subjectivity significantly predict argumentation components. Research findings collectively demonstrate the importance of integrating SSI into science education as an instructional strategy for teaching argumentation and NOS aspects. For this reason, the primary purpose of this study is to contribute to the improvement of NOS understanding and argumentation skills of students by providing a diverse selection of SSI scenarios. These scenarios would cover a wide range of topics that fall under the domains of biology, physics, environmental science, chemistry, and modern ethical dilemmas. By having participants engage in these SSI scenarios, the study aims to create a dynamic learning environment that fosters NOS comprehension and improves the critical thinking skills of students through argumentation. The SSI scenarios presented are relevant to the modern day, preparing students to navigate the complexities of contemporary scientific and societal discourses. As such, the aim of this study was to create scenarios that address SSI to cultivate the teaching of argumentation and SSI.

2.2 Methodology

A methodological approach was employed in the present study to create scenarios that address SSI within the science education context. To generate diverse and engaging SSI scenarios, a comprehensive search was implemented utilizing multiple search engines to reduce bias in the selection of studies. Search engines such as Google Scholar and Shamaa, which publishes articles in Arabic, were leveraged. In this manner, a curated collection of SSI scenarios was gathered, with the intention of providing students with engaging content that would foster debates in the classroom. The aim was to ensure the inclusion of socially relevant and contentious topics. This search process involved examining real-world challenges with ethical implications, such as pollution and the COVID-19 Vaccination. The scenarios cover a range of contexts, including public health, nuclear energy, and environmental impact. The incorporation of these SSI scenarios aimed to enhance student NOS understanding, argumentation skills, and the exploration of the interconnectedness between scientific advancements and broader societal contexts.

Based on the search for SSI, scenarios were created. Following the creation of these scenarios that address SSI, the scenarios were grouped into themes that fit into different science disciplines, which are presented in the next section. After that, the scenarios under the different themes were given to six science teachers (two biology teachers, two chemistry teachers, and two physics teachers) for their content validity. Revisions to the scenarios were made based on the suggestions by the teachers and those were mostly related to language and clarity.

References

Almarashi, A. K., & Alshahri, S. Z. G. (2021). Science, technology, society, and environment (STSE) issues included in the first intermediate grade science course in the Kingdom of Saudi Arabia. 94–67 ,(16)*5* ,المجلة العربية للتربية النوعية. https://doi.org/10.33850/ejev.2021.138595

Alsaaideh, M. H. (2016). The socioscientific issues and sociology of science integrated in the science textbook of the eighth grade and students' understanding level of it. *Studies in Educational Sciences, 43*(Supplement), 1967–1986.

Al-Zunaidi, T. A. R., & Al-Salim, G. H. I. S. (2020). Effectiveness of a program for teaching socio-scientific issues based on an ethical analysis approach on developing reflective thinking skills among high school students in Riyadh. 122–71 ,(24)*4* ,مجلة العلوم التربوية.

BouJaoude, S. (2002). Balance of scientific literacy themes in science curricula: The case of Lebanon. *International Journal of Science Education, 24*, 139–156.

Eastwood, J. L., Sadler, T. D., Zeidler, D. L., Lewis, A., Amiri, L., & Applebaum, S. (2012). Contextualizing nature of science instruction in socioscientific issues. *International Journal of Science Education, 34*(15), 2289–2315.

El Arbid, S. S., & Tairab, H. H. (2020). Science teachers' views about inclusion of socio-scientific issues in UAE science curriculum and teaching. *International Journal of Instruction, 13*(2), 733–748. https://doi.org/10.29333/iji.2020.13250a

Frhod, W., & Abdulwahed, A. A. (2022). Decision making in socioscientific issues of biology teachers for secondary stage. اشراقات تنموية مجلة علمية محكمة *(Illuminations of Development, a Peer-Reviewed Scientific Journal), 34*, 203–228.

Khishfe, R. (2012). Nature of science and decision making. *International Journal of Science Education, 34*(1), 67–100. https://doi.org/10.1080/09500693.2011.559490

Khishfe, R. (2014). Explicit nature of science and argumentation instruction in the context of socioscientific issues: An effect on student learning and transfer. *International Journal of Science Education, 36*(5–6), 974–1016. https://doi.org/10.1080/09500693.2013.832004

Khishfe, R. (2022). Nature of science and argumentation instruction in socioscientific and scientific contexts. *International Journal of Science Education, 44*(4), 647–673. https://doi.org/10.1080/09500693.2022.2050488

Khishfe, R. (2023a). Connected learning: An approach for teaching nature of science aspects and argumentation components. *Journal of Research in Science Teaching, 60*(10), 2180–2215. https://doi.org/10.1002/tea.21855

Khishfe, R. (2023b). Relationship between nature of science and argumentation: A follow-up study. *International Journal of Science and Mathematics Education, 21*(4), 1081–1102. https://doi.org/10.1007/s10763-022-10307-0

Chapter 3
Biology

Abstract This chapter investigates major biological principles in ecology, agricultural science, and public health, focusing on their ethical, environmental, and societal consequences. The first portion looks at biodiversity and biodiversity loss, emphasizing the ongoing conflict between economic growth against environmental protection. It explores biodiversity's ecological, economic, and cultural value, as well as the issues of balancing sustainability and economic growth. The second half discusses agricultural science, including animal production, fish farming, industrial fishing, and developing food technology such as lab-grown meat and vertical farming. This section assesses the environmental impact of food production and the ethical considerations surrounding equitable food distribution. The last section digs into public health challenges, including advances in medicine and biotechnology such as the COVID-19 vaccine, in vitro fertilization, pharmacogenomics, and the ethical quandaries presented by artificial intelligence in healthcare. The chapter offers a balanced discussion of scientific advancement, sustainability difficulties, and ethical considerations, urging a critical outlook on biology's future and position in society.

Keywords Biodiversity conservation · Sustainable agriculture · Food security · Environmental sustainability · Biotechnology and ethics

This chapter is dedicated to socioscientific issues scenarios with an emphasis on three critical areas: Ecology, agriculture, and public health. These scenarios provide a platform for investigating the interaction of societal issues and scientific discovery. In Ecology, biodiversity is investigated, or the diversity of life on Earth, and it becomes a vital lens through which we assess our role in conserving it. In agriculture, ethical and environmental factors take priority, requiring a rethinking of our food production options. Public health scenarios shed light on existing activities, such as responses to COVID-19, while stressing the societal elements of scientific difficulties.

R. Khishfe, *The Complexities of Socioscientific Issues in Education*, Springer Texts in Education, https://doi.org/10.1007/978-3-031-88006-3_3

A. **Ecology**

1. Biodiversity
2. Biodiversity loss

B. **Agricultural science**

3. Livestock Production
4. Equitable Food Production and Distribution
5. The Rise of Fish Farming (Evaluating the Pros and Cons of Fish Farming)
6. Fish Farming (Navigating the Complexities of Fish Farming Ethics and Sustainability)
7. Industrial Fishing
8. Biodiversity and Farming Capacity on Earth
9. Insect Farming
10. Red Meat and Environmental Concerns: A Debate on Consumption and Production
11. Vertical Farming
12. Lab grown meat.
13. The Future of Plant-Based Meat Alternatives
14. Crossbreeding Animals
15. Urban Farming

C. **Public Health**

16. COVID-19 vaccine
17. In Vitro Fertilization
18. Neuralink: A Blessing or a Curse?
19. Pharmacogenomics
20. Endocrine Disruptors and Hormone-Mimicking Chemicals
21. You Got Rid of All the Germs…Now What?
22. Antibiotic Resistance (The Growing Threat to Global Health Security)
23. AI and Medical Ethics
24. Walking a Thin Line: The Debate Surrounding Ozempic

A. Ecology

1. Biodiversity

Biodiversity loss refers to the continued decline of biodiversity, species richness, genetic diversity, and ecosystems. This is a global issue with far-reaching implications. A multifaceted issue that affects multiple sectors, biodiversity loss has ecological, economic, and social implications. While everyone agrees that biodiversity loss is real, the issue's magnitude is contested. In brief, this issue is a tug-of-war between economic gains and wildlife preservation.

One group of people believes that biodiversity loss should be combatted through various ways. After all, this issue goes hand-in-hand with a slew of other problems, such as pollution, overexploitation of natural resources, overfishing, illegal wildlife trade, and climate change. Ways to combat biodiversity loss include weakening consumerist culture, with fewer advertisements and stricter regulations on heavily polluting industries like fast fashion. Squaring off natural reserves would also help endangered species to get their numbers back up.

But detractors argue that prioritizing biodiversity conservation could negatively impact the economy and people's quality of life. They claim that with growing global populations, leaving large areas of land untouched might lead to crowding, poor land management, and increased pollution. Furthermore, they assert that using more land for agriculture and resources is necessary to meet the population's increasing needs. They also argue that promoting reduced consumerism could lead to severe economic losses and significant unemployment. However, this perspective is debated, as numerous studies suggest that biodiversity conservation can yield long-term economic benefits, such as sustaining ecosystem services crucial for agriculture, water purification, and disaster mitigation. Additionally, examples from countries with successful conservation policies show that these efforts can coexist with economic growth through innovations like eco-tourism and sustainable development practices.

The city where you reside is considering implementing a policy to prioritize biodiversity conservation within its boundaries. The decision is put to a vote, and you are asked to express your opinion.

- Would you vote for prioritizing biodiversity conservation in your city?

YES NO

- Explain and justify your decision.
- Another scientist, Professor Ponso, disagrees with your position. How could he explain his position to illustrate the reasons supporting it and convince you?
- What would you reply to Professor Ponso to explain that your position is right?
- How can you explain that scientists reached different conclusions even though they were all looking at the same data about the possible outcomes of biodiversity conservation?
- How do scientists know about the effects of biodiversity conservation?
- How certain are scientists about the knowledge related to biodiversity conservation?
- Do you think the knowledge about biodiversity conservation might change in the future? Explain why or why not.

2. **Biodiversity Loss**

The current issue of biodiversity loss is a global concern that poses a threat to various animal and plant species. In simple terms, biodiversity loss involves the disappearance and extinction of different plants, animals, and other living organisms. In this scenario, two perspectives will be discussed, one in favor of biodiversity loss and the other against it. This is an issue that has been ongoing on a global scale for quite some time.

On the one hand, proponents argue that the loss of biodiversity can have positive implications for society. They provide various examples of why some people believe that this global issue may have some positive aspects for the animal and plant kingdom. Firstly, proponents assert that the loss of biodiversity can open more opportunities for agricultural and economic development, such as land use and job opportunities. They believe that the more physical space available in our environment, the better, regardless of the environmental impact. Secondly, the group proposes that if the animal kingdom has fewer species, it would be easier to control and take care of each species individually. This argument suggests that it would be better for animals if humans could study and understand each species more comprehensively, emphasizing that as a society, we would be better equipped to care for specific species and understand their unique needs on a deeper level.

On the other hand, there are opposing viewpoints. Studies have indicated that the lack of biodiversity in nature can have a detrimental impact on human health. Biodiversity plays a crucial role in preventing the spread of diseases since many species act as natural regulators of diseases. The loss of diversity cuts off human access to several medicinal plants (aloe, turmeric, tulsi, pepper, elaichi, and ginger), important for human health. Moreover, another argument against biodiversity loss emphasizes that it is irreversible. While the lives of animals are in danger due to climate change and insufficient attention to the issue, it is never too late to act. Additionally, this side discusses the idea that plants hold cultural value for individuals in society. With the loss of biodiversity, the uniqueness of humans and past cultures disappears. They claim that the animal and plant kingdom carry stories of the past, and when ecosystems and species vanish from the Earth, everything connected to them also disappears.

- In conclusion, various perspectives have been discussed regarding the loss of biodiversity. Where do you stand? Are you in favor of or against it?
- Explain and justify your position.
- A scientist, Professor Ponso, disagrees with your position. What would he say to make you side with him?
- What would you reply to Professor Ponso to convince him of your position?
- How can you explain that scientists reached different conclusions even though scientists were all looking at the same data about the effects of biodiversity loss?
- How do scientists know about the effects of biodiversity loss?
- How certain are scientists about the knowledge related to biodiversity loss?
- Do you think the knowledge about biodiversity loss might change in the future? Explain why or why not.

B. Agricultural Science
3. Industrial livestock production

Industrial livestock production, encompassing the rearing and breeding of animals for various animal-derived products, has sparked a contentious debate. While some people view industrial livestock production as something that should be done away with, others disagree.

The group that is against industrial livestock production points to its significant environmental impacts and animal welfare concerns. They emphasize issues such as air, soil, and water pollution (Tullo et al., 2019). Greenhouse gas emissions from livestock contribute to air pollution and global warming, with an estimated 14.5% share of total greenhouse gases attributed to this sector (Gerssen-Gondelach et al., 2017), though this figure has been challenged. Manure management practices, deemed inappropriate, are flagged as culprits for soil and water pollution, compounded by the contamination of these resources with antibiotics used in livestock production (Xie et al., 2018). Moreover, practices such as overcrowding, confinement, and restricting access to natural habitats are criticized for their negative impact on animal welfare, with organizations such as the Animal Welfare Committee calling for higher standards in livestock systems.

Conversely, advocates for industrial livestock production counter with an array of mitigation strategies aimed at minimizing adverse environmental effects. World Bank statistics highlight a significant commitment to eco-friendly production, with 61% of livestock financing directed toward such practices (World Bank, 2021). Mitigation strategies involve advanced manure treatment to curtail harmful effects on water and soil, reducing pollution (De Vries et al., 2015). Innovative techniques capturing methane for renewable energy sources are hailed as promising (Lee et al., 2023). Critics of a complete shift away from livestock production argue that it could lead to overuse of water and land resources in extensive agriculture, which often requires more land per unit of food produced (Dorgbetor et al., 2022; Kalyabina et al., 2021).

Advocates also underscore the nutritive aspects of animal-sourced foods, emphasizing their role in providing proteins, amino acids, vitamins, and minerals important for human health, particularly in populations with limited dietary diversity (Leroy & Cofnas, 2020; Steenson & Buttriss, 2020). While these foods are important for addressing global malnutrition, critics question whether they are always "crucial" for health, especially in light of alternative plant-based diets. Furthermore, the economic significance of industrial livestock production is highlighted, providing jobs and income for millions, particularly in low- and middle-income countries (Steinfeld et al., 2006). However, it is debated whether extensive agriculture might employ even more people per unit of food production. Food security is also emphasized, with FAO data asserting that livestock contributes 40% of the global value of agricultural output, supporting the livelihoods and nutrition security of nearly 1.3 billion people (FAO, 2021). Yet critics argue that these benefits are not inherently tied to industrial systems and that alternative practices could address these needs more sustainably.

- Are you with or against industrial livestock production?
- Explain and justify your position.
- Some people disagree with your position. How could they justify their position to convince you and illustrate the reasons supporting their position?
- What would you reply to these people to explain that your position is right?
- How can you explain that scientists reached different conclusions even though scientists were all looking at the same data about the effects of industrial livestock production?
- How do scientists know about the effects of industrial livestock production?
- How certain are scientists about the knowledge related to industrial livestock production?
- Do you think the knowledge about industrial livestock production might change in the future? Explain why or why not.

4. **Equitable food production and distribution**

Food production is vital for human well-being and development, but challenges arise in meeting nutritional demands, particularly with a growing global population.

Severe climatic changes contribute to food shortages, leading to conflicts, wars, and economic migrations. Global issues such as soil degradation, weather changes, population growth, and water security exacerbate the problem, emphasizing the urgency of addressing equitable food distribution (Dos-Santos & Mota, 2019).

A global perspective on food production and distribution reveals alarming statistics. FAO reported over 821 million undernourished people in 2017, with a significant increase in hunger rates from 20.7% to 23.2% between 2014 and 2017 (UN, 2019). African regions face challenges of unequal food distribution, with approximately 237 million undernourished people in 2017. Poor nutrition contributes to 45% of the deaths of children under five, and 22% of the global population is chronically undernourished (UN, 2019). It would be most conducive to tackle world hunger based on a singular, well-connected vision, but there is a debate amongst philanthropists and governments about which approach is best. One of the main points of contention is: Do we have enough food to feed everyone in the first place?

The first group of people believes that we are not currently capable of producing enough food. Thus, we could apply productionism, which is about emphasizing ethical production through technology and innovation (United Nations Environment Programme, 2016). They ascertain that the adoption of advanced technologies, genetically modified crops, and modern farming techniques is necessary to maximize yields and minimize resource use. Advocates rally for substantial investments in the agricultural sect, stating that improved crop yields and expanded production capacity are the cornerstones of ending world hunger. Productionism supporters also highlight the potential for stabilizing food prices and ensuring a reliable food supply, thus reducing vulnerability to climate change and other disruptions, making this system more cost-effective in the long run.

Conversely, another group of people thinks that world hunger is not an issue of resources. The business-as-usual approach relies on current systems (with minor adjustments) to meet global food needs (United Nations Environment Programme, 2016). Its proponents argue that the issue lies in food distribution and accessibility rather than production. By enhancing infrastructure, reducing food waste, and improving market access for smallholder farmers, this approach aims to ensure equitable distribution of existing food resources. This approach's advocates emphasize the importance of maintaining economic and social stability, advocating for gradual improvements within the current framework to achieve food security without disrupting traditional farming practices and the environment.

- In the face of global food distribution challenges and persistent hunger, do you find it more compelling to advocate for a paradigm shift toward local connections between consumers and producers, as suggested by alternative food politics, or to embrace technological solutions and innovation?
- Explain and justify your position.
- Professor Ponso disagrees with your position. What would he say to try and convince you of his viewpoint?
- How would you respond in defense of your original stance?

- How can you explain that scientists reached different conclusions even though scientists were all looking at the same data about food production and distribution?
- How do scientists know about the effects of equitable food production and distribution?
- How certain are scientists about the knowledge related to food production and distribution?
- Do you think the knowledge about equitable food production and distribution might change in the future? Explain why or why not.

5. **The Rise of Fish Farming (Evaluating the Pros and Cons of Fish Farming)**

Fish farming, also known as aquaculture, involves the breeding, rearing, and harvesting of fish and other aquatic organisms for industrial and commercial purposes. This method has gained attention as an alternative to traditional commercial fishing.

Proponents of fish farming argue that it provides a more sustainable and environmentally friendly solution compared to traditional commercial fishing. With many countries heavily reliant on commercial fishing for both protein and economic prosperity, fish farming is seen to meet the demands of a growing population without the adverse consequences of overfishing and damage to wildlife habitats. The aquaculture industry, valued at $78 billion, contributes nearly 40% of the world's consumed seafood and is expanding rapidly (Wolff, 2008). Supporters emphasize that transitioning to fish farming can aid in the recovery of damaged wildlife ecosystems affected by commercial fishing practices. Additionally, fish farming paves the way for economic opportunities and yields more nutritious seafood, according to the National Oceanic and Atmospheric Administration (2019). Overall, this method has seen widespread popularity, with countries like India, Indonesia, and China adopting it (Wee, 2019).

On the other end of the spectrum, detractors of fish farming highlight potential drawbacks associated with this practice. The closed systems used in fish farming increase the risk of parasites and diseases spreading, requiring constant monitoring and regulation. The construction of fish farms may lead to habitat destruction, posing risks to surrounding ecosystems and wildlife. Places like Thailand, Vietnam, and China used to have mangrove forests along their coasts, but some were destroyed to build shrimp and fish farms, as well as other businesses (Wolff, 2008). These mangrove forests were important because they helped blunt the impact of hurricanes, cyclones, and tsunamis. Additionally, there are other negative effects associated with aquaculture, such as escaped farmed fish. Those might interbreed with wild fish, potentially leading to the extinction of hybrid species or competition for mates and resources.

- Do you think that your country would benefit from the use of aquaculture/fish farming?

YES NO

- Explain and justify your position.
- Mrs. Garland disagrees with you. What would she say to try and convince you of her viewpoint?
- How would you respond in defense of your original stance?
- How can you explain that scientists reached different conclusions even though scientists were all looking at the same data about the effects of fish farming?
- How do scientists know about the effects of fish farming?
- How certain are scientists about the knowledge related to fish farming?
- Do you think the knowledge about fish farming might change in the future? Explain why or why not.

6. **Fish Farming (Navigating the Complexities of Fish Farming Ethics and Sustainability)**

Fish farming, also known as aquafarming or aquaculture, involves raising fish and other aquatic creatures in a controlled environment, distinct from commercial fishing that harvests wild fish. While fish farming offers benefits such as reduced pollution compared to commercial fishing, it also has drawbacks.

Proponents of aquaculture state that it is the better option, as commercial fishing contributes to ocean pollution through the loss of synthetic materials and the release of toxic pollutants. Fish farms, as an alternative, can mitigate environmental issues to some extent. Additionally, fish farming generates employment opportunities, particularly in areas with high unemployment rates, potentially improving the financial circumstances of local communities.

However, fish aquaculture has its detractors. It contributes to water contamination as waste products are often dumped into nearby bodies of water, affecting overall water quality. Antibiotic use is another concern in fish farming, where large doses are employed to keep fish healthy in crowded conditions. This practice can lead to antibiotic resistance in humans who consume fish treated with antibiotics, posing significant health risks. The spread of diseases from unsanitary fish farms to consumers is also another significant concern. Additionally, the rapid growth of fish farms might lead to the decline of commercial fishing, which some communities depend on.

- Are you in support of fish farming?

YES NO

- Explain and justify your decision.
- Another scientist, Professor Martina, disagrees with your position. How could she explain her position to illustrate the reasons supporting it and convince you?
- What would you reply to Professor Martina to explain that your position is right?
- How can you explain that scientists reached different conclusions even though scientists were all looking at the same data about the effects of fish farming?
- Do you think the knowledge about fish farming might change in the future? Explain.
- Do you think you might change your decision in the future? Explain.

7. **Industrial Fishing**

Scientists worldwide are engaged in a debate over the impact of fishing on both human health and the environment, weighing the advantages and disadvantages of this practice. The core controversy revolves around whether the focus should be on prioritizing human health or environmental conservation, and which takes precedence.

One faction of scientists contends that fishing is essential because it provides a readily accessible source of one of the most nutritious superfoods. Fish is rich in high-quality protein, omega-3 fatty acids, vitamin D, vitamin B2, calcium, phosphorus, and various minerals like iron, zinc, magnesium, potassium, and iodine. Consuming fish has significant health benefits, including a reduction in blood pressure and the risk of heart attacks or strokes. Given the prevalence of hypertension, particularly in the United States where almost 108 million adults are affected, the nutrients found in fish play a crucial role in promoting and maintaining brain and heart health. Omega-3 fatty acids are essential for health, as our bodies do not produce them naturally. Additionally, fish consumption is seen as a potential solution for health issues like goitre.

On the opposing side, another group of scientists argues that fishing is not only an environmental threat but is actively depleting the populations of key species, such as cod and tuna, at an alarming rate. The decline in aquatic species and the destruction of habitats, especially with industrial-scale fishing, are identified as significant contributors to climate change. Maintaining a robust marine life population is vital for carbon sequestration, a process that helps lock away carbon emissions. The ocean, as the Earth's largest reservoir of carbon, contains approximately 38,000 billion tons of carbon. Scientists warn that rising atmospheric carbon emissions are leading to increased ocean temperatures, impacting aquatic life, and exacerbating the risks posed by mass fishing. The symbiotic relationship between coral and aquatic life is highlighted, emphasizing that damage to one affects the other, further threatening oceanic ecosystems and complicating efforts to reverse human-induced damage. Furthermore, the depletion of fish stocks raises ethical concerns, reflecting a societal failure to protect and sustain the environment. Food security is compromised, particularly in vulnerable communities heavily dependent on natural fishing for sustenance. The contamination of the food supply not only diminishes food safety but also poses a significant risk to the health of communities reliant on fishing as their primary food source.

- As a scientist, do you think fishing should continue?
- Why or why not? Explain.
- Another scientist disagrees with your decision. What points could this scientist make to convince you to side with them?
- What would be your response to the position of the scientist?
- Do you think your decision might change in the future? Why or why not? Explain.
- Why did scientists reach different conclusions even though they were looking at the same data?

8. **Biodiversity and Farming Capacity on Earth**

The challenge at hand revolves around the juxtaposition of limited spaces and an ever-growing demand, particularly concerning the need for vegetation and the prospect of farming on Mars.

Proponents of this idea highlight the potential by pointing to nutrients in the upper crust and the presence of beneficial bacteria and fungi. A study found that certain crops, such as radishes, peas, and tomatoes, can be grown in Mars-like soil on Earth (Wamelink et al., 2014). Thus, with the right technology, scientists might be able to grow these crops on Mars. Additionally, there have been successful experiments with certain viable crops using hydroponics and aeroponics. Hydroponics is the practice of growing plants using a nutrient solution instead of soil. Similarly, aeroponics has to do with cultivating plants in a mist or air setting.

On the contrary, skeptics argue against the feasibility of this endeavor, pointing out the necessity for soil transformation, past failed experiments, and the challenge of managing high pH levels (9.5), with variations in soil treatment further complicating the prospects of Martian agriculture. Moreover, NASA's Curiosity (the Mars rover) highlights the variability of Mars' weather, with extreme temperatures and dust storms at the forefront (Vasavada et al., 2014).

- Considering the challenges and opportunities of farming on Mars, do you believe it is a viable and worthwhile endeavor?

Yes No

- Explain and justify your position.
- An astronaut disagrees with your stance. What would he say to make you change your mind?
- How would you reply to convince him of your perspective?
- Do you think your decision might change in the future? Why or why not? Explain.

- Why did scientists reach different conclusions even though they were looking at the same data?

9. **Insect Farming**

Around the world, insect farming is gaining popularity as an alternative to traditional factory farming, sparking global controversy due to its diverse applications, such as providing an alternative for animal-based protein and animal feed. Sami, a Lebanese expat recently returned from Thailand, is considering opening an insect farm in Lebanon and presents the following proposition:

Scientific evidence and research from Harvard and other U.S. universities demonstrate that insects are a viable alternative to traditional meat (Smith et al., 2021). The studies reveal that incorporating insect-based food into daily diets, particularly in Africa and Asia where insect consumption is common, has effectively addressed deficiencies in protein, zinc, vitamin B12, and anemia. Insects emerge as a valuable nutritional source, particularly in regions facing nutritional shortages. Additionally, insect farming exhibits a lower environmental impact compared to other farming practices like beef farming, which contributes significantly to methane production. Insect farming has a smaller ecological footprint, presenting a more sustainable option. The Food and Agriculture Organization of the United Nations (FAO) asserts that insects are more efficient in converting food into meat, with insects requiring

2 kg of feed to produce 1 kg of insect mass compared to 8 kg of feed needed for cattle (FAO, 2022).

However, this topic is met with the obstacles of cultural aversion and high technological costs. A study by a group called Open Philanthropy (2021) indicates that industry farms often use the same feed for insects as they do for other farm animals, emphasizing the inefficiency of this approach. On top of that, insects can feel and suffer, and the method of slaughter, boiling insects in water vapor, has been deemed inappropriate by the American Veterinary Medical Association.

Sami thinks of opening an insect farm, but wants to take your opinion on whether he will continue with his idea or not:

- Would you advise Sami to open an insect farm?

YES NO

- Explain and justify your decision.
- Professor Smith disagrees with your position. How could he explain his position to illustrate the reasons supporting it and convince you?
- What would you reply to Professor Smith to explain your position is right?
- Why do you think there are different positions from scientists who are studying the same information about insect farming?
- Do you think knowledge about insect farming might change in the future? Explain.
- Do you think you might change your decision in the future? Explain.

10. **Red Meat and Environmental Concerns: A Debate on Consumption and Production**

With growing populations and increasing incomes, the amount of meat consumed per capita has grown (Ritchie et al., 2023). This has led to several implications, most of which are related to the environment. These effects have sparked a debate on whether we should keep consuming meat at the same level as today.

The group that is in favor of not regulating the rates of meat consumption and production argues that red meat is a valuable source of vital nutritional ingredients. It is one of the main sources of vitamin B12, which aids in the making of DNA and in keeping nerve and red blood cells healthy. In addition to that, red meat contains zinc, which ensures the proper functioning of the immune system, and it provides protein, which helps build bones and muscles. The heme iron in red meat is also easily absorbed by the human body, helping to prevent iron deficiency, which is a growing concern, particularly among young women. Additionally, proponents emphasize consumer autonomy-the importance of allowing individuals to choose their dietary preferences-and note that livestock production contributes to the preservation of certain landscapes. Grazing cattle or sheep, for example, can help maintain open grasslands and prevent them from becoming overgrown. Livestock farming also provides non-meat products such as wool and leather, which remain essential in many industries.

The group that argues against high meat consumption, however, is more concerned with the impact of industrial meat production on the environment. The high demand for meat has led to overproduction of livestock, which has multiple negative effects on the environment, including greenhouse gas emissions, animal waste, high water usage, and land degradation. Large areas of land have been deforested and converted into monoculture crop fields solely to feed livestock, destroying natural habitats and reducing biodiversity. These fields are often treated with fertilizers, and manure used as fertilizer releases emissions such as methane, ammonia, and carbon dioxide, which contribute to climate change and water pollution. Manure is typically stored in open lagoons, which can overflow during floods, further polluting waterways. Additionally, livestock themselves emit methane during digestion, a major contributor to climate change. This group advocates for exploring alternative sources of vitamins such as B12, iron, and zinc, and for reducing meat production, which they argue is both feasible and environmentally necessary.

The minister of agriculture is proposing a policy to regulate and significantly reduce the production of livestock, including replanting the land currently used for feeding livestock with other crops. Residents are required to vote on whether to support or oppose this policy.

- Would you vote for or against this issue?
- Explain and justify your decision.
- Dr. Saliba, an environmental specialist, disagrees with your position. How could you explain her position to illustrate the reasons supporting it and convince you?
- What would your response be to Dr. Saliba?
- How do you think your position will affect the environment in the future?
- Do you think your decision might change in the future? Why or why not? Explain.

- Why did scientists reach different conclusions even though they were looking at the same data?

11. **Vertical Farming**

Environmentalists widely advocate for the implementation of vertical farms, asserting that it can alleviate hunger amidst the challenges posed by climate change, including increased dryness and heat. Facing unprecedented disasters such as wildfires, which threaten global food supplies, the need for innovative production methods becomes urgent. Vertical farming, defined as the practice of growing stacked produce in a closed and controlled environment, emerges as an innovative solution, garnering support from many.

One key advantage of vertical farming lies in its ability to optimize space compared to traditional farming methods. By utilizing shelves stacked one above the other, it achieves increased production in a limited space, ensuring more homogeneous harvests and reducing food waste. The method allows for multiple harvests of the same product throughout the year, providing a reliable means of production. The controlled environment, involving artificial temperature regulation, light control, and gas management, shields crop from the impacts of climate change, ensuring consistent yields. Additionally, vertical farming reduces pollution by minimizing the use of pesticides, herbicides, and nitrogen fertilizers, safeguarding soil and groundwater. This approach not only benefits the environment but also offers health advantages to consumers, providing fresh, high-quality, and pesticide-free produce.

However, the reliance on vertical farming comes with environmental and societal consequences. Opponents argue that the closed and regulated environment obstructs the entry of insects and disrupts the process of pollination. Furthermore, plants grown in this environment may have fewer nutrients than their outdoor counterparts, even if they are labeled as organic. While vertical farming is considered reliable, its dependence on various technologies for lighting, temperature, and humidity introduces vulnerabilities. A single power outage could result in significant losses, posing risks for consumers relying on these crops. Importantly, the shift to vertical farming has implications for communities heavily dependent on traditional agriculture, potentially destabilizing their economic foundations. Many farmers who sustain their livelihoods from the land may face poverty as this new technology transforms agricultural practices.

Your city plans on relying on vertical farming to provide food for its residents.

- Would you add for reliant on vertical farming?

Yes No

- Explain and justify your decision.
- Another scientist, Professor Dupont, disagrees with your position. How could he explain his position to illustrate the reason supporting it and convince you?
- What would you reply to Professor Dupont to explain your position is right?
- How can you explain that scientists reached different conclusions even though scientists were all looking at the same data about effects of vertical farming?

- Do you think the knowledge about vertical farming might change in the future? Please explain.
- Do you think you might change your decision in the future? Please explain.

12. **Lab-Grown Meat**

Lab-grown meat, also known as cultured or cultivated meat, is a technological innovation revolutionizing the food industry. It involves producing animal meat from animal cells in a controlled laboratory environment. Despite recent FDA approvals for lab-grown meat production and consumption, skepticism and criticism persist regarding environmental, nutritional, and health claims.

Proponents argue that lab-grown meat offers potential environmental benefits, with studies indicating that greenhouse gas emissions could be reduced by over 95% compared to traditional meat production, provided that renewable energy sources power production systems. Currently, the traditional animal agriculture industry is responsible for 14.5% of global greenhouse gas emissions (University of California Riverside, 2023). Additionally, the controlled environment in lab-grown meat production minimizes contamination risks from pathogens such as salmonella, often found in traditional slaughter processes. Lab-grown meat is typically free of growth hormones and antibiotics, which proponents argue promotes consumer health. Advocates also highlight potential economic advantages, suggesting that lab-grown meat could become cost-competitive as production scales up, eliminating many challenges associated with raising livestock, such as disease outbreaks in confined animal farming.

Critics, however, raise concerns about lab-grown meat's feasibility and broader implications. They argue that the energy-intensive nature of bioreactors could lead to significant greenhouse gas emissions unless production techniques rely primarily on renewable energy sources, which may not yet be practical or cost-effective in many regions (University of California Riverside, 2023). Health concerns are also central to their critique, particularly regarding unknown long-term risks from the cellular multiplication processes used to grow the meat. Additionally, they question the "naturalness" of lab-grown meat and its affordability for the average consumer, advocating for more extensive research before it becomes widely available.

From an ethical perspective, some critics question whether eliminating traditional animal farming could inadvertently affect animal populations, ecosystems, and biodiversity. For example, traditional livestock farming practices often create grazing land that supports certain species. They also emphasize that the product may not resonate with consumers seeking minimally processed, natural foods. The executive director of the Food Ethics Council has argued that lab-grown meat, while innovative, may ultimately be perceived as just another processed food (Mello-Klein, 2022).

Given these perspectives, the Ministry of Agriculture is proposing to support research and development of lab-grown meat while seeking public input on its potential role in the food industry.

- If you were asked to vote on whether the government should heavily invest in scaling lab-grown meat production, what would your decision be?
- Explain and justify your stance.
- Dr. Saliba, an environmental specialist, disagrees with your position. How could she explain her position and convince you?
- How would you defend your stance against Dr. Saliba's?
- How can you explain that scientists reach different conclusions about the environmental and health impacts of lab-grown meat even though they analyze the same data?
- How do scientists know about the effects of lab-grown meat?
- How certain are scientists about the knowledge related to lab-grown meat?
- Do you think our understanding of lab-grown meat will evolve in the future? Explain why or why not.

13. **The Future of Plant-Based Meat Alternatives**

The transition toward plant-based diets with reduced meat consumption is widely regarded as crucial for mitigating the detrimental impacts of the food system on the planet while improving human health and animal welfare.

Advocates argue that reducing meat intake is essential for addressing environmental concerns and promoting healthier dietary habits. One proposed strategy is the adoption of plant-based meat alternatives (PBMAs), which aim to mimic the sensory attributes of animal meat, such as taste, texture, and appearance. Proponents emphasize that PBMAs can significantly reduce greenhouse gas emissions, water use, and land requirements compared to traditional livestock farming, while also eliminating the ethical concerns associated with animal suffering.

Critics, however, question whether PBMAs truly provide a sustainable and healthy alternative. They point out that many PBMAs have long ingredient lists, including additives, and differ nutritionally from animal meat. For example, some PBMAs contain less protein or higher sodium levels. Furthermore, critics raise concerns about the environmental impact of producing PBMAs, particularly regarding the sustainability of sourcing key ingredients like soy or the energy-intensive processing methods.

Additionally, skeptics argue that PBMAs are not always affordable or accessible to all consumers and may not appeal to those who prioritize natural, unprocessed foods. While PBMAs are marketed as a solution for reducing meat consumption, some believe alternative approaches, such as promoting whole plant foods or improving sustainable livestock practices, may be more effective.

Society faces a key question: Should plant-based meat alternatives play a central role in the transition to sustainable diets, or should we focus on other strategies?

- Are you in favor of or against the widespread adoption of plant-based meat alternatives?
- Explain and justify your position.
- Some people may disagree with your stance. How could they justify their position and illustrate the reasons supporting it?
- What would your response be to defend your position?
- Why might scientists reach different conclusions about the impacts of plant-based meat alternatives even when analyzing the same data?
- How do scientists gather knowledge about the effects of plant-based meat alternatives?
- How certain are scientists about the benefits and drawbacks of plant-based meat alternatives?
- Do you think our understanding of plant-based meat alternatives will evolve in the future? Why or why not?

14. **Crossbreeding Animals**

Crossbreeding is the process of mating two animals from different breeds within the same species to produce offspring with desirable traits (aesthetic, behavioral, or biological). This practice is highly debated due to its ethical, economic, and biological consequences.

Supporters of crossbreeding emphasize the concept of *hybrid vigor* or *heterosis,* which describes the appearance of healthier or more desirable traits in offspring. This can lead to benefits such as improved immunity, reduced disease risks, and economic advantages. For instance, Boxer dogs are prone to brachycephalic airway syndrome due to their short facial structure (Argos, 2022). Crossing them with a longer-faced breed may reduce this risk. In agriculture, crossbreeding has proven advantageous, as hybrid cows have better feed efficiency. Studies found that crossbreds are 12% more efficient than pure Holstein cows while weighing 50 kilograms heavier on average (VikingGenetics, 2023). Additionally, crossbreeding allows certain animals to convert feed into body mass more efficiently, which reduces environmental strain from livestock farming.

On an individual level, crossbreeding can also produce aesthetically appealing traits or desirable behavioral traits. For example, *labradoodles*—the cross between poodles and Labradors—function as effective service and therapy dogs. Poodles are known for being high-strung and prone to certain health issues, while Labradors can trigger allergic reactions in some people (Massey, 2022; Dugal, 2024). This cross offers a balance, offsetting risks associated with each breed.

Opponents, however, argue that crossbreeding raises ethical concerns and places profit over animal welfare. They maintain that businesses exploit animals, focusing only on profit instead of ensuring the animals' quality of life. Additionally, critics claim that crossbreeding can lead to health risks for offspring, such as genetic defects and susceptibility to certain diseases. For instance, *puggles* (crosses between pugs and beagles) often suffer from breathing issues due to pinched nostrils (Howson, 2019). Furthermore, crossbreeds like mules, resulting from horse and donkey breeding, face reproductive challenges as they are often infertile (Rodriguez, 2007). Critics also highlight that many animals are crossbred solely for aesthetic traits, with no proven health benefits for the offspring (Howson, 2019).

Your local government is considering introducing policies to decrease crossbreeding practices in animal breeding within the agricultural and companion animal industries. This would involve regulatory oversight, stricter breeding guidelines, and possibly public education campaigns. Residents must vote on whether they support or oppose these proposed changes.

- Do you vote for or against the proposed changes?
- Explain and justify your decision.
- Another expert, Professor Youssef, disagrees with your position. What does he say to explain his stance?
- How would you respond to Professor Youssef to defend your original viewpoint?
- What scientific studies or evidence would help determine the risks and benefits of crossbreeding?
- How certain are scientists about the data surrounding crossbreeding's effects?
- Do you think public opinion or scientific understanding about crossbreeding might change in the future? Why or why not?

15. **Urban Farming**

As cities become more densely populated and technologically advanced, urban farming is gaining attention as a potential solution to food security and climate change challenges. Urban farming includes a variety of practices, such as rooftop gardens, community gardens, and small-scale farms within city limits (University of Michigan Maize & Blue Cupboard, 2021). These initiatives are being promoted by many as sustainable strategies to address food security, reduce the effects of climate change, and support urban health. However, critics point to financial costs, environmental concerns, and technological challenges that question their long-term feasibility.

Supporters of urban farming emphasize its environmental and technological benefits. These farms incorporate sustainable methods such as water harvesting, composting, and integrated pest management to conserve resources and reduce waste (Hydropolis, 2022). They are also thought to minimize the use of pesticides and fertilizers by using innovative techniques and creating opportunities for recycling waste into sustainable food systems. Additionally, urban farms are positioned near city populations, reducing transportation distances for crops, cutting pollution, and ensuring city residents have fresher produce (Hydropolis, 2022; Arnst, 2024).

On the other hand, opponents raise concerns about urban farming's challenges and limitations. Establishing urban farms requires significant financial investments in infrastructure and technology (Bardsley et al., 2024). Furthermore, urban farms depend on large amounts of energy for practices like artificial lighting, nutrient solutions, and climate control. In fact, producing just one kilogram of urban farming crops requires 38.8 kWh, compared to only 5.4 kWh for traditional rural farming (Labrie & Stride, 2023). Critics argue that these higher energy demands, combined with costs, limit accessibility and scalability, especially in lower-income areas. Additionally, some urban farms are only capable of growing leafy greens, strawberries, and mushrooms, making them insufficient as a long-term or large-scale solution (Bardsley et al., 2024).

Your local government is proposing the relocation of some agricultural production from rural farmland into urban landscapes. The goal is to repurpose this land for urban housing projects while integrating urban farming to meet food supply needs. However, this plan raises environmental concerns, financial hurdles, and social implications. The community will vote on whether or not to approve this proposal.

- Do you vote for or against this plan?
- Explain and justify your decision.
- Another expert, Professor Yara, disagrees with your position. What does she say to illustrate the reasons behind her stance and convince you?
- What would you reply to Professor Yara to explain that your position is right?
- How certain are scientists about the knowledge related to urban farming?
- Do you think the knowledge about urban farming might change in the future? Explain why or why not.
- Do you think you might change your decision in the future? Explain.

C. Public Health
16. **COVID-19 Vaccination**

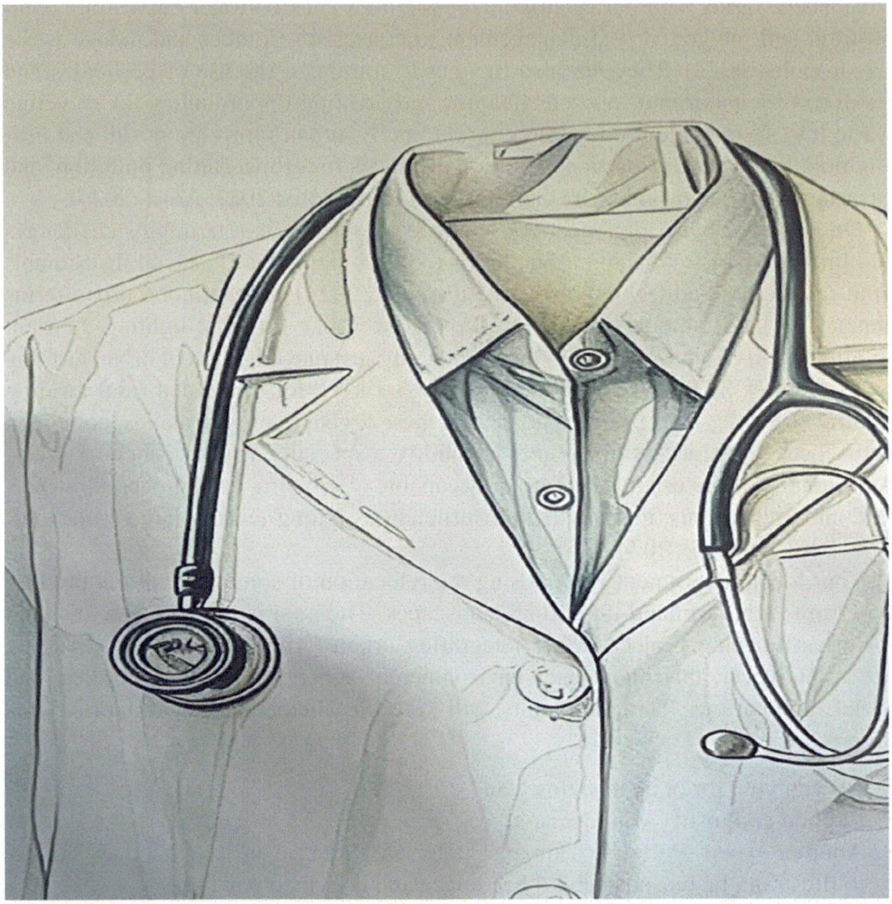

Coronavirus disease (COVID-19) is an infectious disease causing severe respiratory illnesses, and scientific evidence shows that it is caused by the SARS-CoV-2 virus (Johns Hopkins Medicine, 2022). COVID-19 has led to the deaths of millions of people globally across varying ages and is also believed to leave lasting and severe symptoms for many infected individuals after recovery. Due to the widespread contagion and the tremendous negative effects it has had on populations globally, the spread of the Coronavirus was declared a worldwide pandemic. To control the pandemic, many establishments and institutions began mandating or requiring vaccines. This means that all individuals in a certain country or at a certain institution must be vaccinated to work, travel, get an education on campuses, and even attend some social gatherings. Being among these institutions, the American University of Beirut (AUB) [a university located in Lebanon] is requiring all its staff and students to get vaccinated for entry to the Beirut campus. This vaccination mandate has brought about many controversies along the way, with large groups of people either agreeing or disagreeing with the vaccine mandate.

Individuals arguing for the vaccine mandate take a scientific standpoint and believe that it is necessary to vaccinate all students and personnel at AUB to ensure their safety and health. The vaccination campaign at AUB has stated that its priority is providing a safe campus for everyone and reducing the death toll and infection spread within the population at AUB that would affect the external population of Lebanon. Those in support of the vaccine mandate believe that requiring students and staff to get the vaccine is the only way to minimize the spread of the virus and protect others (American University of Beirut, 2020). Those arguing for the vaccine mandate also believe that the Coronavirus is very dangerous to individuals vulnerable to respiratory illnesses and of an older age, meaning everyone on the AUB campus has a duty to protect these individuals by getting the vaccine. Another significant aspect of the vaccination campaign at AUB and the supporters of the vaccine argue that education needs to be provided in a comfortable campus environment to optimize learning, which can only be achieved by implementing the vaccine mandate forcefully.

On the other hand, the group against the implementation of the AUB vaccine mandate argues that it is unethical to force a vaccine on people, and the vaccine might have unknown negative side effects or implications. These people believe that the vaccine's ingredients are still unknown, which might lead to negative side effects in the future due to insufficient research on the topic, and its fabrication was done quickly. Individuals against the vaccine mandate at AUB also feel that the vaccine is imposed on them, since rejecting it would lead to consequences such as the inability to acquire an education. Moreover, the group against the vaccine mandate argues that they do not feel the need to get the vaccine since many have already contracted and recovered from COVID-19, deeming vaccination as unnecessary and avoidable.

- Do you support the COVID-19 Vaccination Mandate at AUB?
- Explain and justify your decision.
- Did you get the COVID-19 Vaccine? Why or why not?
- A fellow scientist, Dr. Margaretta Alfonso, has argued in support of the COVID-19 vaccination mandate at AUB since she believes it is the only way to encourage students to protect themselves and others from catching the COVID-19 disease. What would her argument be?
- How would Dr. Alfonso's point affect your position?
- Do you think your position will change in the future? Explain.
- Why did scientists reach different conclusions even though they were looking at the same data?

17. In Vitro Fertilization and Designer Babies

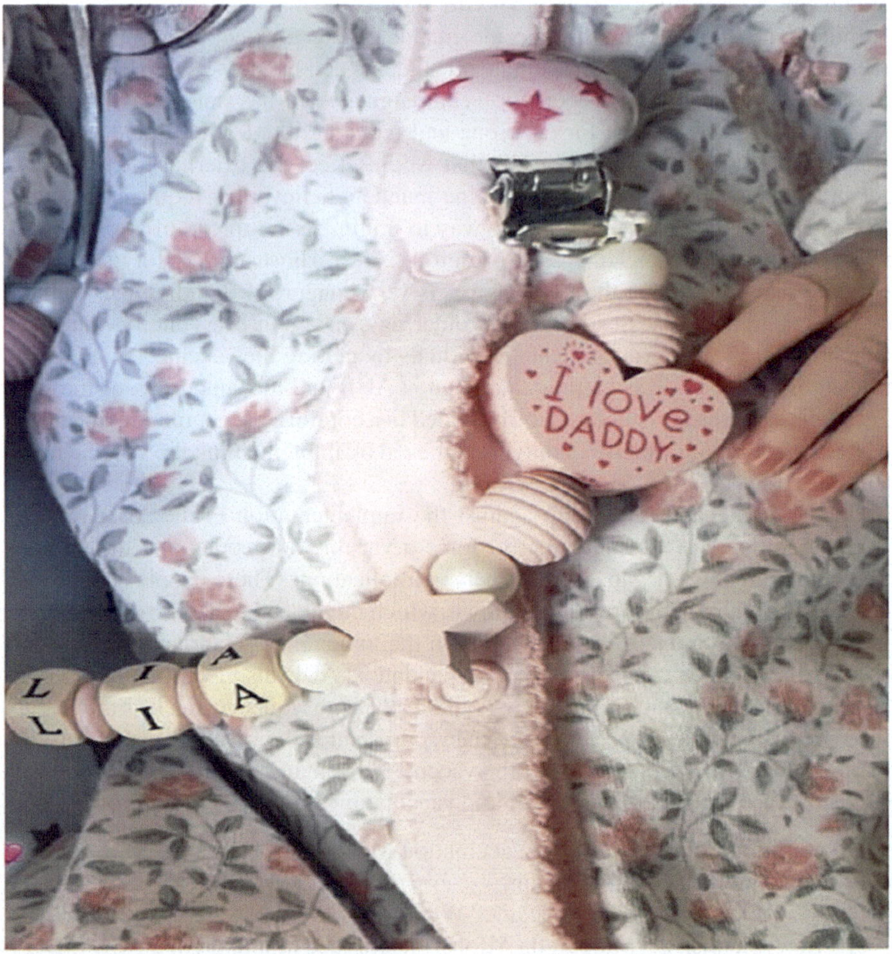

In Vitro Fertilization (IVF) is a form of assisted reproductive technology wherein an egg is extracted from the ovaries and fertilized with sperm in a controlled medical facility setting. The fertilized egg, now called the embryo, is then implanted in the woman's womb for further development. This procedure can use a woman's own eggs and her partner's sperm, or in some cases, involve donors for eggs and/or sperm, or even, in countries where this is permitted, the use of a surrogate to carry the pregnancy. Due to the high cost, IVF is typically a last resort after more affordable infertility treatments have proven unsuccessful. In the realm of reproductive medicine, a combination of preimplantation genetic testing/diagnosis (PGT/D) and in vitro fertilization is employed to prevent genetic diseases in embryos. The reasons for these alterations have evolved with technology. Genetically altered embryos can be formed by introducing desired genetic material into the embryo, sperm, or egg cells of the parents. This could potentially be achieved in the future by directly depositing favored genes into the cell or by using gene-editing technology, a process often associated with the term 'designer babies.' A designer baby is a fetus genetically engineered in vitro for carefully selected traits, potentially reducing the risk of common diseases, or allowing for gender selection and could be further tailored for specific cosmetic traits or skills such as musical giftedness. Preimplantation genetic diagnosis enables the screening of viable embryos for certain harmful or unfavorable traits within the genes before implantation into the mother. This technique is often employed when parents carry genes that may pose risks to their offspring.

According to a BBC article featuring biochemist Dr. Smith, the technology behind genetically modified babies is considered a promising advancement in addressing severe medical conditions, such as extending healthy human lifespan; however, Dr. Smith also emphasizes the need to address ethical concerns and ensure public approval of such innovations.

An alternative argument in support of designer baby technology is that parents would have control over potential outcomes for their children's lives. This perspective contends that ethical issues related to genetic selection should be diminished as parents routinely make decisions that shape a child's development. However, critics argue that this perspective overlooks significant ethical concerns, including the risk of creating social inequalities and economic divisions, as access to this technology could be limited to those who can afford it. Additionally, another perspective on this socioscientific issue persists, with varying views on when the use of genetically modifying embryos through IVF is ethical. The Council on Ethical and Judicial Affairs (CEJA), a body of the American Medical Association (AMA) in the United States, released a statement in the late 1990s allowing genetic selection to prevent or cure diseases but considered selection based on benign characteristics as largely unethical. Scientists opposing designer babies raise concerns about the moral and societal implications of genetically modified children with selected desirable traits. The argument against designer babies from a moral standpoint revolves around the fear that if this technology becomes normalized, it could lead to divisions between those who can afford the technology and those who cannot. This could lead to economic divisions becoming genetic divisions, raising unpredictable ethical implications in the future.

- Do you believe that the use of IVF, in combination with preimplantation genetic testing and genetic modification, is an ethically justifiable practice?
- Explain and justify your position.
- Doctor Bay disagrees with your position. What would she say to convince you of her position?
- What would you say in defense of your original stance?
- How can you explain that scientists reached different conclusions even though scientists were all looking at the same data about the effects of IVF?
- How do scientists know about the effects of IVF?
- How certain are scientists about the knowledge related to IVF?

- Do you think the knowledge about IVF might change in the future? Explain why or why not.

18. **Neuralink: A Blessing or a Curse?**

Neuralink, founded and led by Elon Musk, is a company that aims to develop a generalized brain-computer interface that can restore autonomy and unlock human potential. Mainly, Neuralink hopes to use its microchips to help people with certain medical conditions. The way the chips work is by interpreting signals produced in the brain and relaying information to devices via Bluetooth. The development and imminent distribution of Neuralink's microchips has sparked some debate.

Proponents argue that Neuralink has the potential to revolutionize human-machine interaction, providing solutions for neurological disorders, injuries, and cognitive limitations. The Neuralink chip is purported to eventually be able to treat conditions such as paralysis and blindness. Additionally, Neuralink supporters envision broader applications, such as improving memory, cognitive processing, and even enabling direct communication between individuals. This group believes that Neuralink is a safe and ethical way to augment human intelligence and creativity, and that it will foster a more harmonious and inclusive society.

While everyone can agree that those goals sound noble, critics are concerned with the logistics and trustworthiness of the whole ordeal. They argue that this technology poses serious risks to the environment, health, and human dignity. This group warns that Neuralink could cause irreversible damage to the brain, increase the vulnerability to cyberattacks and manipulation, and create a digital divide between those who can afford the devices and those who cannot. They worry about whether these technologies might be covered by insurance. This group also questions the moral implications of altering the natural state of the human mind, and fears that Neuralink will erode the values and identity of humanity. This group opposes the development and use of Neuralink devices and calls for strict regulation and oversight.

You are an American citizen. The state government has put Neuralink's development to a referendum, and the decision will determine whether to halt or speed up Neuralink's efforts based on the result.

- What do you vote for?
- Explain and justify your decision. Someone said:
- A sci-fi enthusiast disagrees with you. What do they say to convince you of their position?
- What do you say back, in defense of your position?
- Do you think your decision might change in the future? Why or why not? Explain.
- Why did scientists reach different conclusions even though they were looking at the same data?

19. **Pharmacogenomics**

Pharmacogenomics, a facet of personalized medicine, tailors drug treatments (pharmaco) based on individual genetic factors (genomics). Recognizing that genetic differences influence how the body responds to drugs, pharmacogenomics aims to optimize treatment outcomes by aligning medications with patients' specific genetic profiles.

Proponents of pharmacogenomics argue that it expedites the treatment process by selecting the most suitable medication for individuals. Given that about 90% of drugs exhibit optimal efficacy in only around 50% of patients, pharmacogenomics holds the promise of improving drug responses (Pirmohamed, 2023). Supporters emphasize the potential to prevent thousands of deaths annually attributed to adverse drug reactions. Additionally, pharmacogenomics allows for personalized dosages, recognizing that individuals may metabolize drugs differently. It has demonstrated success in commercial applications, particularly in the treatment of cancers and blood coagulation disorders (Vogenberg et al., 2010).

But critics of pharmacogenomics argue that its strength is also its weakness. Drawing up a genetic profile for a patient can be time-consuming and ultimately futile, as reactions to drugs are not just dictated by one gene but involve an interplay between an arsenal of genes. Moreover, opponents highlight the financial barriers, suggesting that pharmacogenomics is costlier than conventional medical approaches and may not be covered by medical insurance, potentially creating disparities, especially in developing areas (Miao et al., 2020). Legal concerns are raised, as genetic testing involves multiple parties, posing the risk of misusing genetic information and leading to issues such as blackmail, invasions of privacy, and social stigmatization.

You are a doctor. You attend a conference on pharmacogenomics, and you start thinking about whether to apply the new information you have just learned.

- What do you decide to do?
- Explain and justify your decision.
- A pharmacist disagrees with you. What do they say to convince you of their position?
- What do you say back, in defense of your position?
- How can you explain that scientists reached different conclusions even though scientists were all looking at the same data about the effects of pharmacogenomics?
- How do scientists know about the effects of pharmacogenomics?
- How certain are scientists about the knowledge related to pharmacogenomics?
- Do you think the knowledge about pharmacogenomics might change in the future? Explain why or why not.

20. **Endocrine Disruptors and Hormone-Mimicking Chemicals**

Nowadays, we are perpetually surrounded by chemicals we are unaware of. Some of these can have specific effects on us, as they might affect the way our bodies produce hormones that regulate all our bodily functions, from our temperature to our focus. These chemicals are referred to as endocrine disruptors (because they disrupt the function of the endocrine hormone system) or hormone-mimicking chemicals (because they might trick the body into behaving as it does when they secrete a certain hormone). They are present in some cosmetics, soaps, shampoos, plastics, and even air fresheners (Dodson et al., 2012). Their ubiquity has made some unafraid of them, while others have become only more alarmed by their obtrusive presence.

Advocates for the regulation of endocrine disruptors and hormone-mimicking chemicals argue that proactive measures are essential to protect public health and the environment. They point to mounting scientific evidence linking exposure to these chemicals to a range of adverse health outcomes, including reproductive disorders (Laws et al., 2021), neurodevelopmental effects (Schug et al., 2015), and disruptions to the endocrine system (Diamanti-Kandarakis et al., 2009). Additionally, they emphasize the precautionary principle, which suggests that in the absence of conclusive evidence, preventive action should be taken to avoid potential harm. Proponents advocate for stringent regulations, including bans or restrictions on the use of certain chemicals in consumer products, as well as comprehensive testing and monitoring programs to assess exposure levels and evaluate potential risks.

Critics of regulations targeting endocrine disruptors and hormone-mimicking chemicals raise concerns about the scientific uncertainty surrounding their health effects and the potential economic consequences of stringent regulations. They argue that while some studies suggest a correlation between chemical exposure and adverse health outcomes, the causal relationship remains unclear, and further research is needed to establish definitive conclusions. Moreover, they contend that overly restrictive regulations could impose significant costs on industries and consumers, leading to job losses, reduced innovation, and higher prices for goods and services. Additionally, they question the feasibility and effectiveness of regulatory approaches, suggesting that voluntary industry initiatives and consumer education campaigns may be more appropriate and cost-effective strategies for managing chemical risks.

You are a food manufacturer. Your team uses Bisphenol A to make food containers, which mimics estrogen and has thus been linked to a myriad of diseases, from reproductive disorders to developmental abnormalities. The FDA, on the other hand, asserts that it is safe (U.S. Food and Drug Administration, 2023).

- Do you keep using Bisphenol A?
- Explain the reasoning behind your answer.
- A food scientist, Ed Ebbel, disagrees with your decision. What would he say to convince you of his position?
- How would you defend your original stance?
- Do you think your decision might change in the future? Why or why not? Explain.
- Why did scientists reach different conclusions even though they were looking at the same data?

21. **You Got Rid of All the Germs...Now What?**

In recent years, the issue of antimicrobial resistance (AMR) has garnered increasing attention due to its potential to undermine modern medicine's ability to combat infectious diseases effectively. Among the various contributing factors to this crisis are the overuse and misuse of chemical disinfectants. Chemical disinfectants, while important for maintaining hygiene, contribute to the development of antimicrobial resistance.

On one side of the debate are individuals who do not think this issue is significant. This perspective often stems from (a) a lack of awareness or (b) poor understanding of the mechanisms driving antimicrobial resistance. Many people perceive chemical disinfectants as a quick and effective solution to kill harmful pathogens (due to insistent advertising), without fully grasping the broader implications of their widespread use. Additionally, with our fast-paced daily lives, some may prioritize immediate results over long-term consequences, opting for the ease of using chemical disinfectants without considering their potential role in fueling antimicrobial resistance. As previously stated, the marketing of these products often emphasizes their potency and efficacy in eliminating germs, overshadowing discussions about their impact on microbial resistance. The focus is on the noxiousness of germs, not how our bodies react to germs.

Those who are aware of how the human immune system might react decry this overreliance on chemical disinfectants. This group includes scientists, public health experts, and environmental advocates who warn against the indiscriminate use of these substances. Their opposition is grounded in scientific evidence highlighting the role of chemical disinfectants in selecting for resistant microbial strains (Rozman et al., 2021). Studies have demonstrated that repeated exposure to disinfectants can promote the development of resistance mechanisms in bacteria, rendering them less susceptible to antimicrobial agents. Furthermore, the widespread dissemination of antimicrobial-resistant pathogens in healthcare settings and the environment underscores the urgent need for alternative strategies to mitigate AMR. Advocates for reducing chemical disinfectant use advocate for approaches such as promoting hand hygiene, implementing infection control measures, and investing in research for novel antimicrobial agents that minimize the risk of resistance development.

- Which group do you agree with?
- Explain and justify your decision.
- Mary, your neighbor who has three kids, disagrees with you. What does she say to convince you of her position?
- How would you reply in defense of your stance?
- Do you think your decision might change in the future? Why or why not? Explain.
- Why did scientists reach different conclusions even though they were looking at the same data?

22. **Antibiotic Resistance (The Growing Threat to Global Health Security)**

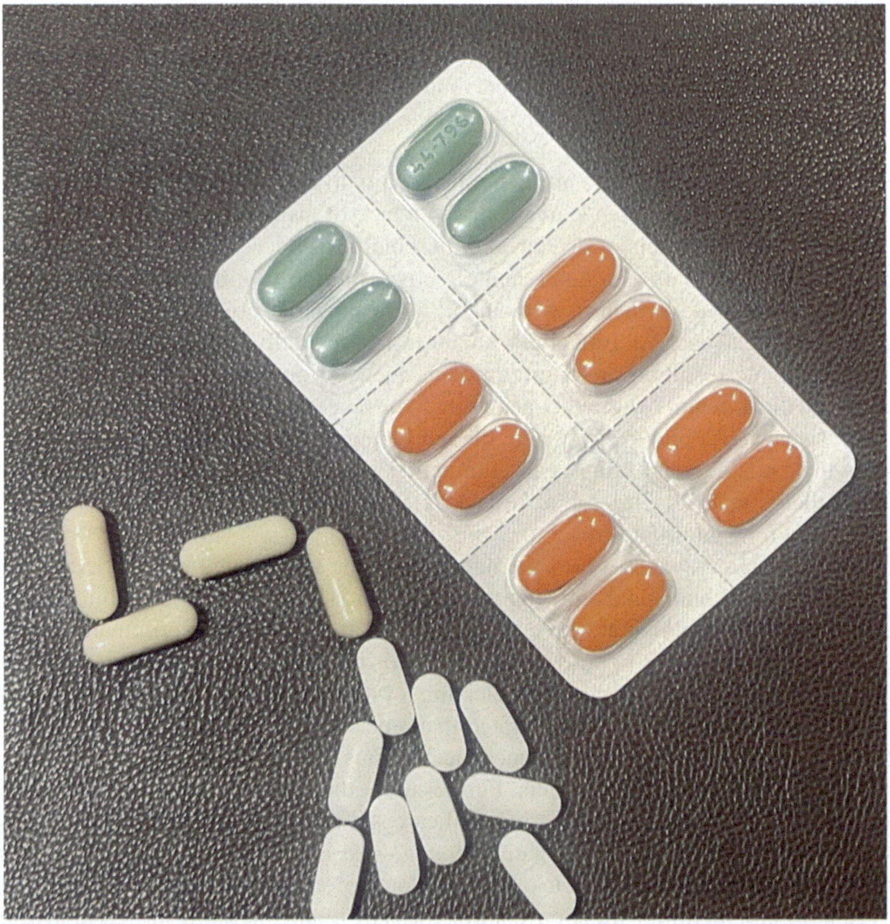

Antibiotic resistance poses a danger to global health security, reminiscent of the days when even small illnesses might be lethal. The overuse and abuse of antibiotics has compounded the situation by creating a breeding environment for resistant microorganisms.

One group claim that urgent action is necessary to stem the emergence of resistance, including prudent antibiotic use, infection prevention and control, and a cohesive strategy across the human and animal health sectors. However, while the necessity of combating antibiotic resistance is apparent, there remain obstacles to overcome.

One argument against forceful intervention contends that economic incentives in the pharmaceutical sector prevent investment in novel antibiotics, resulting in a standstill of innovation. Indeed, present economic models may unwittingly discourage researchers from exploring novel alternatives. Furthermore, the difficulty of coordinating global efforts, as well as a lack of strong surveillance data in low- and middle-income nations, provide considerable impediments to successful action.

Stewardship programs, which are coordinated strategies aimed at optimizing the use of antibiotics to combat resistance, have been introduced globally. These programs focus on educating healthcare providers and the public about appropriate antibiotic use, preventing overprescription, and monitoring antibiotic consumption. The goal is to reduce unnecessary antibiotic use while ensuring that essential medicines remain available for those who truly need them.

- Should governments intervene more in pharmaceuticals to spur antibiotic development, or trust market forces? How to balance innovation needs with concerns about government control?
- Is it ethical to prioritize antibiotic research over other health issues, or does it neglect broader health equity? How to justify resources allocation amid competing health priorities?
- Are global collaborations the best way to tackle antibiotic resistance, or should countries focus locally? How to balance global coordination with sovereignty concerns?
- Should stewardship programs focus on cutting overall antibiotic use or specifically target inappropriate use? How can we balance reducing resistance with ensuring adequate access to antibiotics for those who need them?
- Do you think your decision might change in the future? Why or why not? Explain.
- Why did scientists reach different conclusions even though they were looking at the same data?

23. **AI and Medical Ethics**

Artificial intelligence (AI) in healthcare is a rapidly advancing field that offers significant benefits, such as improved diagnostic accuracy and personalized treatment plans. These advancements could streamline medical decisions, save patients time and money, and optimize resource usage in hospitals. However, AI also raises ethical concerns, including algorithmic biases, lack of transparency, and risks to sensitive patient data. Research reveals strong opinions both for and against the adoption of AI in medical practices, highlighting the complex ethical dilemmas involved.

Groups supporting the use of AI in healthcare emphasize its potential to improve patient outcomes and diagnostic precision. AI technologies, particularly deep learning models, can analyze vast amounts of medical data faster and more accurately than human doctors, enabling early disease detection and personalized treatment strategies (Johnson et al., 2020). According to Andrés Lawson, AI systems can process medical images and identify early-stage cancers at a lower cost and higher speed compared to conventional methods (Yoon & Amadiegwu, 2023). Additionally, AI reduces administrative burdens on healthcare workers by managing and organizing patient data, allowing medical staff to focus on critical and complex cases

(Johnson et al., 2020). Nirav R. Shah, a physician at Stanford School of Medicine, highlights this issue: "Healthcare is not productive. …As a doctor, admitting a patient takes me an hour and 45 minutes because I'm a glorified data entry clerk" (Columbus, 2024). Many respected institutions, such as Johns Hopkins Medicine, Mayo Clinic, the World Health Organization, and Yale New Haven Health, endorse integrating AI into healthcare systems, predicting it will improve efficiency, reduce unnecessary tests, and lower costs (Coalition for Health AI, 2024; World Health Organization, n.d.).

Groups opposed to AI in healthcare raise critical concerns about its limitations and risks. AI systems can reinforce existing biases within the medical data they are trained on, leading to inequitable and inaccurate outcomes, particularly for underserved populations (Levi, 2023). For example, if data for a small demographic group is insufficient, AI could consistently misdiagnose members of that group while still achieving high overall accuracy rates (Horowitz, 2024). Critics also argue that AI systems often operate as "black boxes," meaning they produce conclusions without revealing the reasoning or processes behind their decisions. This lack of transparency makes it challenging for healthcare workers to trust and validate AI-driven recommendations, hindering accountability and decision-making (Johnson et al., 2020). Furthermore, the integration of AI requires handling sensitive patient data, raising concerns about data security and privacy breaches that could impact patients' lives, including employment or social opportunities (Murdoch, 2021).

City officials are exploring ways to encourage local hospitals to adopt AI technologies for diagnostics, treatment planning, and data management. Residents are being asked to vote on whether to support this initiative.

(a) What do you vote for?
(b) Explain and justify your decision.
(c) Another expert, Professor Youssef, disagrees with your position. What does he say to illustrate the reasons behind his stance and convince you?
(d) What would you reply to Professor Youssef to explain that your position is right?
(e) Why did scientists reach different conclusions even though they were looking at the same data?
(f) Do you think the knowledge about AI and medical ethics might change in the future? Explain.
(g) Do you think you might change your decision in the future? Explain.

24. **Walking a Thin Line: The Debate Surrounding Ozempic**

Ozempic, also known as semaglutide, is a medication approved by the FDA to treat type 2 diabetes. It works by helping the pancreas produce more insulin, which lowers blood sugar levels. Additionally, the drug suppresses cravings and increases feelings of fullness, resulting in weight loss for many patients (Beckett, 2023; Cleveland Clinic, 2024). Recently, Ozempic has gained widespread attention for its off-label use as a weight loss drug, sparking debates about its benefits, risks, and appropriate use.

Supporters of Ozempic as a weight loss medication argue that obesity is a complex, chronic condition that requires medical intervention. They contend that lifestyle changes like diet and exercise alone are often not sufficient. Research shows that combining Ozempic with healthy lifestyle changes leads to significant weight loss: participants lost an average of 34 pounds over 68 weeks, while those who relied solely on lifestyle changes lost only 6 pounds on average (Wilding et al., 2021). Proponents also highlight the additional health benefits of weight loss with Ozempic, including reduced blood pressure, cholesterol, and blood sugar levels—factors often associated with obesity (Cleveland Clinic, 2024). Furthermore, semaglutide, Ozempic's active ingredient, has already been approved by the FDA at a higher dose under the name *Wegovy* for chronic weight management.

Critics, however, raise concerns about Ozempic's safety and unintended consequences. They point to common side effects, such as nausea, vomiting, diarrhea, and constipation, as well as more severe risks like pancreatitis, gallbladder disease, and thyroid cancer (Buchler, 2024). Rapid weight loss caused by the medication can result in muscle loss, alongside the so-called "Ozempic face," where loss of facial fat accelerates signs of aging (Suran, 2023). Critics also note that once patients stop using Ozempic, their hormones return to normal, and the lost weight is often regained (Suran, 2023; Cleveland Clinic, 2024). Socially, the drug's popularity among celebrities and influencers has led to shortages for diabetes patients and misuse by individuals who are not obese but want "cosmetic" weight loss. Finally, while Ozempic has FDA approval for diabetes treatment, its off-label use for weight loss raises ethical concerns, as its long-term effects remain unclear (Northrop, 2024; Buchler, 2024).

Your city is considering allowing healthcare providers to prescribe Ozempic as an anti-obesity medication.

- What do you vote for?
- Explain and justify your decision.
- Another expert, Professor Youssef, disagrees with your position. What does he say to illustrate the reasons behind his stance and convince you?
- What would you reply to Professor Youssef to explain that your position is right?
- How do scientists know about the effects of ozempic?
- How certain are scientists about the knowledge related to ozempic?
- Do you think the knowledge about ozempic might change in the future? Explain why or why not.

References

American University of Beirut. (2020). AUBe Vaccinated Campaign: Towards a safer AUB community. American University of Beirut. https://www.aub.edu.lb/emergency/AVC/Pages/default.aspx?utm_source=chatgpt.com

Argos. (2022, September 2). *Do pedigree dogs have more health problems than crossbreeds?* Argos Pet Insurance. Retrieved from https://www.argospetinsurance.co.uk/we-talk-pet/are-pedigree-dogs-more-likely-to-have-health-problems-than-crossbreeds/

Arnst, J. (2024, March 20). *Free, weekly fresh produce improved diet, physical activity, and reduced CVD risk factors*. American Heart Association Newsroom. Retrieved from https://newsroom.heart.org/news/free-weekly-fresh-produce-improved-diet-physical-activity-and-reduced-cvd-risk-factors

Bardsley, D. K., Winsborough, S., Skinner, W., & Drew, G. (2024). The governance of hydrosocial risk in peri-urban South Australia. *Geographical Research, 62*(4), 553–568.

Beckett, E. (2023, March 19). *Ozempic helps weight loss by making you feel full. But certain foods can do the same thing – Without the side effects*. The Conversation. Retrieved from https://theconversation.com/ozempic-helps-weight-loss-by-making-you-feel-full-but-certain-foods-can-do-the-same-thing-without-the-side-effects-201870#:~:text=The%20active%20ingredient%20in%20Ozempic,helps%20manage%20type%202%20diabetes

Buchler, E. G. (2024, January 12). *The pros, cons, and unknowns of popular weight-loss drugs*. Johns Hopkins University Hub. Retrieved from https://hub.jhu.edu/2024/01/11/ozempic-wegovy-weight-loss-drugs-pros-cons/

Cleveland Clinic. (2024, July 10). *Ozempic for weight loss: Who should try it and will it work?* Cleveland Clinic. Retrieved from https://health.clevelandclinic.org/ozempic-for-weight-loss

Coalition for Health AI. (2024). *Introduction page*. Retrieved July 15, 2024, from https://www.coalitionforhealthai.org/

Columbus, L. (2024, July 12). *Five ways generative AI is improving healthcare today and defining its future*. VentureBeat. Retrieved from https://venturebeat.com/ai/five-ways-generative-ai-is-improving-healthcare-today-and-defining-its-future/

De Vries, J. W., Groenestein, C. M., Schröder, J. J., Hoogmoed, W. B., Sukkel, W., Koerkamp, P. G., & De Boer, I. J. M. (2015). Integrated manure management to reduce environmental impact: II. Environmental impact assessment of strategies. *Agricultural Systems, 138*, 88–99.

Diamanti-Kandarakis, E., Bourguignon, J. P., Giudice, L. C., Hauser, R., Prins, G. S., Soto, A. M., Zoeller, R. T., & Gore, A. C. (2009). Endocrine-disrupting chemicals: An Endocrine Society scientific statement. *Endocrine Reviews, 30*(4), 293–342. https://doi.org/10.1210/er.2009-0002

Dodson, R. E., Nishioka, M., Standley, L. J., Perovich, L. J., Brody, J. G., & Rudel, R. A. (2012). Endocrine disruptors and asthma-associated chemicals in consumer products. *Environmental Health Perspectives, 120*(7), 935–943. https://doi.org/10.1289/ehp.1104052

Dorgbetor, I. K., Ondrasek, G., Kutnjak, H., & Mikuš, O. (2022). What if the world went vegan? A review of the impact on natural resources, climate change, and economies. *Agriculture, 12*(10), 1518.

Dos-Santos, M. L., & Mota, M. (2019). Toward sustainable and smart cities in Africa: A review and challenges. In C. Guedes & C. Cantuaria (Eds.), *Bioclimatic architecture in warm climates* (pp. 299–309). Springer Nature.

Dugal, G. (2024, July 16). *Poodle vs Labradoodle: The differences (with pictures)*. Dogster. Retrieved from https://www.dogster.com/dog-breeds/poodle-vs-labradoodle#:~:text=Both%20breeds%20are%20extremely%20friendly,better%20for%20new%20dog%20owners

Food and Agriculture Organization of the United Nations. (2019). The state of food security and nutrition in the world 2019: Safeguarding against economic slowdowns and downturns (1st ed.). United Nations.

Food and Agriculture Organization of the United Nations. (2021). Animal production. https://www.fao.org/animalproduction/en.

Gerssen-Gondelach, S. J., Wicke, B., & Faaij, A. P. (2017). GHG emissions and other environmental impacts of indirect land use change mitigation. *GCB Bioenergy, 9*(4), 725–742.

Horowitz, B. T. (2024, March 27). *How AI bias is impacting healthcare*. InformationWeek. Retrieved from https://www.informationweek.com/machine-learning-ai/how-ai-bias-is-impacting-healthcare#close-modal

Howson, E. (2019, November 3). *Cross-breeding: Blessing or curse?* Redbrick. Retrieved from https://www.redbrick.me/cross-breeding-blessing-or-curse/

Hydropolis. (2022, October 11). *Vertical farming as a new type of urban farming*. Hydropolis. Retrieved from https://www.hydro-polis.com/en/vertical-farming-as-a-new-type-of-urban-farming/.

Johns Hopkins Medicine. (2022). COVID-19. https://www.hopkinsmedicine.org/health/conditions-anddiseases/coronavirus.

Johnson, K. B., Wei, W., Weeraratne, D., Frisse, M. E., Misulis, K., Rhee, K., Zhao, J., & Snowdon, J. L. (2020). Precision medicine, AI, and the future of personalized health care. *Clinical and Translational Science, 14*(1), 86–93. https://doi.org/10.1111/cts.12884

Kalyabina, V. P., Esimbekova, E. N., Kopylova, K. V., & Kratasyuk, V. A. (2021). Pesticides: Formulants, distribution pathways and effects on human health – A review. *Toxicology Reports, 8*, 1179–1192.

Labrie, P., & Stride, A. (2023, January 26). *Energy and underground farming*. GreenForges. Retrieved from https://www.greenforges.com/blog/energy-and-underground-farming#:~:text=According%20to%20the%202021%20Global,Lighting%20%2D%2055%25

Laws, M. J., Neff, A. M., Brehm, E., Warner, G. R., & Flaws, J. A. (2021). Endocrine disrupting chemicals and reproductive disorders in women, men, and animal models. *Advances in Pharmacology (San Diego, Calif.), 92*, 151–190. https://doi.org/10.1016/bs.apha.2021.03.008

Lee, J., Smith, A., & Patel, R. (2023). Methane capture, utilization, and sequestration technology based on novel approaches. *Chemical Engineering Journal, 450*, 138967. https://doi.org/10.1016/j.cej.2023.138967

Leroy, F., & Cofnas, N. (2020). Should dietary guidelines recommend low red meat intake? *Critical Reviews in Food Science and Nutrition, 60*(16), 2763–2772.

Levi, J. (2023, June 6). *AI in medicine needs to be carefully deployed to counter bias – and not entrench it*. National Public Radio. Retrieved from https://www.npr.org/sections/health-shots/2023/06/06/1180314219/artificial-intelligence-racial-bias-health-care

Massey, T. (2022, February 16). *Labradoodle vs Labrador*. Snowy Pines White Labs. Retrieved from https://www.snowypineswhitelabs.com/blog/labradoodle-vs-labrador/#:~:text=If%20you're%20allergic%20to,go%20wrong%20with%20the%20Labrador

Mello-Klein, C. (2022, November 21). *Could lab-grown meat pave the way for more ethical, environmentally friendly food?* Northeastern Global News. Retrieved from https://news.northeastern.edu/2022/11/21/lab-grown-meat/

Miao, L., Zhang, J., Yi, L. X., & Huang, S. (2020). The ethical, legal, and regulatory issues associated with pharmacogenomics. In W. Cai, Z. Liu, L. Miao, & X. Xiang (Eds.), *Pharmacogenomics in precision medicine: From a perspective of ethnic differences* (pp. 219–239). Springer Singapore. https://doi.org/10.1007/978-981-15-3895-7_12

Murdoch, B. (2021). Privacy and artificial intelligence: Challenges for protecting health information in a new era. *BioMed Central Medical Ethics, 22*. https://doi.org/10.1186/s12910-021-00687-3

National Oceanic and Atmospheric Administration. (2019, April 2). *What is aquaculture?* National Oceanic and Atmospheric Administration's National Ocean Service. Retrieved from https://oceanservice.noaa.gov/facts/aquaculture.html

Northrop Grumman. (2024, March 19). How Northrop Grumman employees can manage healthcare cost increases. https://www.theretirementgroup.com/featured-article/5448095/how-northrop-grumman-employees-can-managehealthcare-cost-increases

Pirmohamed, M. (2023). Pharmacogenomics: Current status and future perspectives. *Nature Reviews Genetics, 24*(6), 350–362. https://doi.org/10.1038/s41576-022-00572-8

Ritchie, H., Rosado, P., & Roser, M. (2023, December). *Meat and dairy production*. Our World in Data. Retrieved May 22, 2024, from https://ourworldindata.org/meat-production

Rodriguez, M. (2007, June 20). *Why can't mules breed?* The Tech Interactive. Retrieved from https://www.thetech.org/ask-a-geneticist/articles/2007/ask225/

Rozman, U., Pušnik, M., Kmetec, S., Duh, D., & Šostar Turk, S. (2021). Reduced susceptibility and increased resistance of bacteria against disinfectants: A systematic review. *Microorganisms, 9*(12), 2550.

Schug, T. T., Blawas, A. M., Gray, K., Heindel, J. J., & Lawler, C. P. (2015). Elucidating the links between endocrine disruptors and neurodevelopment. *Endocrinology, 156*(6), 1941–1951. https://doi.org/10.1210/en.2014-1734

Smith, M. R., Stull, V. J., Patz, J. A., & Myers, S. S. (2021). Nutritional and environmental benefits of increasing insect consumption in Africa and Asia. *Environmental Research Letters, 16*(6), 065001. https://doi.org/10.1088/1748-9326/abf06c

Steinfeld, H., & Livestock, Environment and Development (Firm), & Food and Agriculture Organization of the United Nations. (2006). *Livestock's long shadow: Environmental issues and options*. Food and Agriculture Organization of the United Nations.

Steenson, S., & Buttriss, J. L. (2020). Healthier and more sustainable diets: What changes are needed in highincome countries? *Nutrition Bulletin, 46*(3), 279–309. https://doi.org/10.1111/nbu.12518

Suran, M. (2023, April 26). As Ozempic's popularity soars, here's what to know about semaglutide and weight loss. *JAMA Network*. Retrieved from https://jamanetwork.com/journals/jama/article-abstract/2804462

Tullo, E., Finzi, A., & Guarino, M. (2019). Review: Environmental impact of livestock farming and precision livestock farming as a mitigation strategy. *The Science of the Total Environment, 650*(Pt 2), 2751–2760. https://doi.org/10.1016/j.scitotenv.2018.10.018

U.S. Food and Drug Administration. (2023, April 20). *Bisphenol A (BPA): Use in food contact application*. Retrieved April 13, 2024, from https://www.fda.gov/food/food-packaging-other-substances-come-contact-food-information-consumers/bisphenol-bpa-use-food-contact-application

United Nations Environment Programme. (2016). *Food systems and natural resources*. Retrieved from https://www.resourcepanel.org/reports/food-systems-and-natural-resources

United Nations Food and Agriculture Organization. (2022). *The state of food security and nutrition in the world 2022*. Retrieved from https://www.fao.org/documents/card/en?details=cc0639en

University of California Riverside. (2023). *Lab-grown meat's carbon footprint potentially worse than retail beef*. News Bites Pty Ltd.

University of Michigan Maize & Blue Cupboard. (2021, April 23). *Urban farming: The introduction of vertical farming*. Retrieved from https://mbc.studentlife.umich.edu/2021/04/23/urban-farming-the-introduction-of-vertical-farming/

Vasavada, A. R., Grotzinger, J. P., Arvidson, R. E., Calef, F. J., Crisp, J. A., Gupta, S., Hurowitz, J., Mangold, N., Maurice, S., Schmidt, M. E., Wiens, R. C., Williams, R. M. E., & Yingst, R. A. (2014). Overview of the Mars Science Laboratory mission: Bradbury Landing to Yellowknife Bay and beyond. *Journal of Geophysical Research: Planets, 119*(6), 1134–1161. https://doi.org/10.1002/2014JE004622

VikingGenetics. (2023, March 20). *Crossbreeding improves feed efficiency*. Retrieved from https://www.vikinggenetics.com/news/crossbreeding-improves-feed-efficiency#:~:text=A%20new%20study%20conducted%20by,the%20environment%20and%20animal%20welfare

Vogenberg, F. R., Barash, C. I., & Pursel, M. (2010). Personalized medicine: Part 1: Evolution and development into theranostics. *Pharmacy and Therapeutics, 35*(10), 560–576. Retrieved from https://www.researchgate.net/publication/47636594_Personalized_Medicine_Part_1_Evolution_and_Development_into_Theranostics

Wamelink, G. W., Frissel, J. Y., Krijnen, W. H. J., Verwoert, M. R., & Goedhart, P. W. (2014). Can plants grow on Mars and the Moon: A growth experiment on Mars and Moon soil simulants. *PLoS One, 9*(8), e103138. https://doi.org/10.1371/journal.pone.0103138

Wee, R. Y. (2019, March 18). *Top countries for aquaculture production*. WorldAtlas. Retrieved from https://www.worldatlas.com/articles/top-15-countries-for-aquaculture-production.html

Wilding, J. P. H., Batterham, R. L., Calanna, S., Davies, M., Van Gaal, L. F., Lingvay, I., McGowan, B. M., Rosenstock, J., Tran, M. T. D., Wadden, T. A., Wharton, S., Yokote, K., Zeuthen, N.,

Kushner, R. F., & STEP 1 Study Group. (2021). Once-weekly semaglutide in adults with overweight or obesity. *The New England Journal of Medicine, 384*(11), 989–1002. https://doi.org/10.1056/NEJMoa2032183
Wolff, A. (2008, August 4). *The pros and cons of fish farming*. Encyclopædia Britannica. Retrieved from https://www.britannica.com/explore/savingearth/the-pros-and-cons-of-fish-farming
World: Moving towards sustainability: The livestock sector and the World Bank. (2021). Asia News Monitor.
World Health Organization. (n.d.). *Harnessing artificial intelligence for health.* Retrieved July 15, 2024, from https://www.who.int/teams/digital-health-and-innovation/harnessing-artificial-intelligence-for-health
Xie, W. Y., Shen, Q., & Zhao, F. J. (2018). Antibiotics and antibiotic resistance from animal manures to soil: A review. *European Journal of Soil Science, 69*(1), 181–195.
Yoon, S., & Amadiegwu, A. (2023, June 21). *Emerging tech, like AI, is poised to make healthcare more accurate, accessible and sustainable*. World Economic Forum. Retrieved from https://www.weforum.org/agenda/2023/06/emerging-tech-like-ai-are-poised-to-make-healthcare-more-accurate-accessible-and-sustainable/

Chapter 4
Physics

Abstract This chapter delves into crucial current physics subjects such as renewable and nuclear energy, quantum computing breakthroughs, gravitational waves, and space exploration ethics. It investigates the advantages and disadvantages of solar electricity, notably in Lebanon, focusing on environmental issues and economic viability. The debate over nuclear energy balances its role in combating climate change against dangers such as radioactive waste and potential accidents. The ramifications of quantum computing on cybersecurity and encryption are also discussed, emphasizing both the revolutionary potential and the security risks posed by this technology. Furthermore, the chapter looks at the relevance of gravitational waves, wondering whether sustained investment in astrophysics research is necessary. Lastly, the argument over space exploration and tourism considers both the potential of scientific development and the ethical issues of environmental sustainability and resource allocation. The chapter fosters scientific research while recognizing the complexity of technology advancement and its societal consequences.

Keywords Renewable energy · Nuclear energy · Quantum computing · Environmental impact · Gravitational waves

This chapter looks at socioscientific scenarios in the field of physics, with an emphasis on two crucial topics: solar panels and nuclear energy. These scenarios are about more than simply understanding the science content; they also raise crucial societal implications. First, there is an investigation of how solar panels use sunlight to create electricity and examine their role in sustainable energy. Subsequently, there is an exploration of nuclear energy, unraveling its mechanics and discussing the continuing argument over whether we should phase it out.

1. Solar Panels
2. Solar Power in Lebanon
3. Nuclear Energy
4. Should Nuclear Energy Be Phased Out of Use?
5. Should We Fear Quantum Computing?

R. Khishfe, *The Complexities of Socioscientific Issues in Education*, Springer Texts in Education, https://doi.org/10.1007/978-3-031-88006-3_4

6. Gravitational Waves: How Much Science Is Too Much?
7. Should We Reach for the Stars?
8. Particle Therapy

1. Solar Panels

Solar energy is the technology used to harness the sun's energy and make it usable. It employs photovoltaic cells, commonly referred to as solar panels. These cells are composed of semiconductor materials like those found in computer chips. When these cells are exposed to sunlight, it dislodges electrons from their atoms, causing the electrons to flow through the cell, generating electricity.

The utilization of solar power offers several benefits. Firstly, it provides financial advantages. By adopting solar energy, residential energy consumers can achieve significant savings on their home energy bills. This transition reduces dependence on conventional electricity and gas services, which is especially beneficial given the substantial increases in electricity and utility gas service prices in recent years (Jackson, 2023). In fact, solar-powered electricity is almost 30% cheaper than the cheapest form of fossil-fuel-generated electricity (Hill, 2023). Furthermore, the adoption of solar energy carries environmental benefits. It plays a significant role in reducing fossil fuel emissions. Even in regions with less sunlight, the use of solar panels can lead to substantial reductions in emissions. For instance, a typical home in Connecticut, a state not particularly known for its sunny weather, using solar panels can achieve emission reductions equivalent to planting about 150 trees annually over the system's lifetime (Energy Sage, 2021). Meanwhile, fossil fuel alternatives cause the emission of carbon dioxide, sulfur dioxide, and sulfur dioxide, which causes acid rain.

However, it is essential to acknowledge that the process of generating electricity from solar energy, while not directly impacting global warming emissions, is associated with emissions in other stages of the solar life cycle, such as manufacturing, installation, materials transportation, maintenance, and dismantlement. Estimates indicate that life-cycle emissions for photovoltaic systems typically range from 0.07 to 0.18 pounds of carbon dioxide equivalent per kilowatt-hour. The manufacturing of PV cells involves the use of hazardous materials, many of which are employed for cleaning and purifying the semiconductor surface. These chemicals include hydrochloric acid, sulfuric acid, nitric acid, hydrogen fluoride, 1,1,1-trichloroethane, and acetone. Workers in the industry also face risks related to the inhalation of silicon dust (Union of Concerned Scientists, 2013). Moreover, solar panels degrade and become progressively less efficient over time, often requiring replacement after about 25–30 years. While they are being promoted worldwide as a key tool in reducing carbon emissions, the disposal, replacement, and recycling of billions of old solar panels pose environmental challenges. On the other hand, burning fossil fuels is more efficient and cost-effective in the short term because of the fossil-fuel-dependent systems set up by most governments. There is currently a lack of infrastructure to effectively handle the waste generated by aging solar panels, potentially leading to a significant waste issue by 2050 (Gordon, 2023). "Unless we establish recycling chains now," warns Ute Collier, deputy director of the International Renewable Energy Agency, "it's going to be a waste mountain by 2050. We're producing more and more solar panels, which is great, but we must address how we'll manage the waste" (Gordon, 2023).

- Considering the financial advantages and potential environmental challenges associated with solar energy, do you believe the long-term benefits of reduced energy bills and fossil fuel emissions outweigh the environmental concerns related to the manufacturing, lifespan, and recycling of solar panels?

Yes No

- Explain and justify your position.
- The mayor of a mid-sized American city (e.g., Electra City) disagrees with your position. What would he say to convince you of his position?
- What would you say back, in defense of your original position?
- How can you explain that scientists reached different conclusions even though scientists were all looking at the same data about the effects of solar panels?
- How do scientists know about the effects of solar panels?
- How certain are scientists about the knowledge related to solar panels?
- Do you think the knowledge about solar panels might change in the future? Explain why or why not.

2. Solar Power in Lebanon

Viewing Lebanon from above, you would think it was a first-world country. Ironically, the economic crisis has turned the average Lebanese person's roof into a glistening jewelry box of metal and plastic. As citizens turn to solar panels to combat economic crises and high electricity costs, questions arise about the long-term sustainability and potential environmental consequences of this renewable energy source.

Opponents to this position express concerns about the environmental implications of solar panels, particularly when they reach the end of their operational life. Solar panels consist of various materials, some of which are toxic, leading to challenges in responsible disposal and recycling. According to cancer biologist David H. Nguyen, PhD, toxic chemicals in solar panels include copper indium gallium (di) selenide, lead, and polyvinyl fluoride. Critics emphasize the potential risks associated with the presence of hazardous components, especially in regions where recycling infrastructure is not widespread.

Supporters argue that solar panels are an eco-friendly solution for generating electricity, emphasizing their role in reducing greenhouse gas emissions and lessening dependence on fossil fuels. Proponents highlight the renewable nature of solar energy, presenting it as a crucial step toward a more sustainable and environmentally conscious energy infrastructure. Additionally, organizations like the RCRA effectively recycle depleted solar panels, evading the risk of pollution.

You are a Lebanese citizen. An environmental organization is surveying people in regard to their opinions.

- Which side do you agree with?
- Explain and justify your position.
- The mayor of Electra City disagrees with your position. What would he say to convince you of his position?
- What would you say back, in defense of your original position?
- Do you think your decision might change in the future? Why or why not? Explain.
- Why did scientists reach different conclusions even though they were looking at the same data?

3. Nuclear energy

Nuclear energy refers to the energy produced through nuclear reactions, specifically nuclear fission, or fusion. It is used to generate electricity in nuclear power plants, which has led to controversy.

One group of scientists supports the use of nuclear energy to generate electricity due to the advantages it offers. First, it produces small amounts of greenhouse gases, as it is a low-carbon energy source. Nuclear power facilities produce baseload power, which can complement intermittent renewable energy sources. They provide a constant and reliable supply of electricity. Additionally, nuclear energy has a high energy density, allowing it to yield vast amounts of electricity from small amounts of fuel (Elhegazy & Kamal, 2021).

On the other hand, some scientists are against the use of nuclear energy due to the negative consequences that come with it. Some of the challenges include the disposal and safe management of radioactive waste, the fear of the spread of nuclear weapon technology, potential safety risks due to meltdowns or mistakes, and public mistrust. Furthermore, nuclear accidents could lead to environmental and social disasters, as seen in the prominent cases of Chernobyl and Fukushima. Additionally, it is important to note that sustaining nuclear power plants and managing nuclear waste is highly expensive (Rhodes & Beller, 2000). Moreover, the development of nuclear energy raises concerns about the potential for dual-use technology, as the same technology and materials used in civilian nuclear power generation can also be used to produce nuclear weapons. This creates risks related to the spread of nuclear weapon technology and international security.

Countries always seek to balance the benefits and risks of nuclear energy to produce the energy they need while managing the safety and environmental consequences associated with it.

Your government plans to build nuclear power plants to produce electricity and provide it to the country and asks its citizens to vote for or against this decision.

- Would you vote in favor of constructing nuclear power plants in your country?

YES NO

- Explain and justify your decision.
- Another scientist, Professor Ponso, disagrees with your position. How could he explain his position to illustrate the reasons supporting it and convince you?
- How would you respond to Professor Ponso to explain that your position is correct?
- Do you think your decision might change in the future? Why or why not? Explain.
- Why did scientists reach different conclusions even though they were looking at the same data?

4. Should Nuclear Energy Be Phased Out of Use?

Nuclear energy, derived from the core of an atom through nuclear fission, where atoms are split, has been a significant source of low-carbon power since the 1950s, utilized in over 50 countries. The debate surrounding nuclear energy revolves around its potential benefits and drawbacks.

Supporters argue that it is a clean and reliable source of power, complementing renewables and aiding in decarbonization. One Olympic-sized swimming pool of nuclear fuel is enough to power more than 140 million homes for a year (Office of Nuclear Energy, 2022). It produces minimal waste compared to traditional fossil fuels, but the waste it does generate is highly hazardous and requires careful long-term management. Nuclear waste, such as spent fuel and other byproducts, remains radioactive for thousands of years and must be securely stored to avoid environmental and human health risks. Some countries, like France and Japan, recycle portions of their nuclear waste to reduce its long-term environmental impact, but the management and disposal of waste remain a significant challenge (Office of Nuclear Energy, 2022). It requires less space, and, despite initial high costs, can be efficient for small-scale use. Even nuclear waste can partially be recycled, with countries like France and Japan doing so. Economically, nuclear energy is seen as a crucial complement to renewables for full decarbonization.

On the contrary, opponents emphasize the environmental risks and the challenges associated with both the routine management of radioactive waste and the potential for catastrophic nuclear accidents, such as those seen in Fukushima and Chernobyl. They argue that the long-lasting nature of nuclear waste makes its routine storage and handling hazardous, even without accidents, and that this requires significant financial and logistical resources to manage safely. On top of that, they claim that even if we were to double the number of nuclear plants by 2050, we would only decrease greenhouse gases by 4% (Leman, 2022). The United States, for instance, treats nuclear waste as particularly volatile, limiting its use by using uranium and plutonium only once, unlike France and Japan. The urgency of addressing climate change is stressed by those against nuclear energy, highlighting the slow and expensive nature of nuclear plant construction compared to faster, renewable alternatives. The call is for directing subsidies to sustainable energies, signaling a shift away from nuclear power in favor of a renewable future.

- Do you believe nuclear power should continue to be a major part of our energy mix?

Yes No

- Defend your position by providing arguments related to environmental impact, safety concerns, efficiency, and the urgency of addressing climate change.
- Professor Ponso disagrees with you. What would he say to try and convince you otherwise?
- How would you defend your original stance?
- Do you think your decision might change in the future? Why or why not? Explain.
- Why did scientists reach different conclusions even though they were looking at the same data?

5. Should We Fear Quantum Computing?

The two goals of computers are the following: accuracy, and speed. Quantum computing has taken the latter to new heights. In brief, this type of computing uses the complex behavior of subatomic particles such as protons and electrons to perform calculations. As a result, it can consider several different possibilities in one step, unlike traditional classical computing, culminating in significantly faster processes. The advancement of this technology has ignited extensive discourse within the realm of cybersecurity, particularly concerning its potential impact on encryption—the cornerstone of online privacy. Quantum computers, with their potential for unparalleled speed and processing power, pose a formidable threat to existing encryption methods, leading to apprehensions about the vulnerability of sensitive data transmitted over the internet. However, amidst concerns about national security and the potential ramifications of quantum computing, there exists a faction that heralds its revolutionary capabilities.

Proponents of quantum computing seek to temper apprehensions by emphasizing the timeline for its practical implementation and its potential societal benefits. Some experts suggest that quantum computers capable of breaching current encryption standards are still at least 15 years away from mass production (Burgan, 2023). Moreover, they highlight the myriad ways in which quantum computing stands to revolutionize various sectors, including healthcare, infrastructure development, and environmental conservation. For instance, quantum computing could expedite drug discovery processes, optimize construction projects, and facilitate the reduction of greenhouse gas emissions (Brownell, 2019; Davies, 2023). By accentuating these potential benefits, advocates aim to assuage concerns and foster a more balanced discourse surrounding quantum computing.

On the other end, opponents of quantum computing emphasize the grave implications it may have for national security. With the ability to effortlessly decipher passwords, quantum computers could expose corporate trade secrets and compromise bank data. In other words, quantum computers could precipitate widespread job losses, economic upheaval, and personal financial ruin. Notably, many detractors are industry experts, including entities like the U.S. National Security Agency (NSA), which has advocated for preemptive measures to fortify encryption standards and secure infrastructure against the looming threat of quantum computing (U.S. National Security Agency, 2023). Such concerns underscore the urgency of adopting proactive strategies to mitigate potential risks.

- Do you believe we should slow down the development of quantum computing?

Yes No

- Defend your position by providing arguments.
- Professor Ponso disagrees with you. What would he say to try and convince you otherwise?
- How would you defend your original stance?
- How can you explain that scientists reached different conclusions even though scientists were all looking at the same data about the effects of quantum computing?
- How do scientists know about the effects of quantum computing?
- How certain are scientists about the knowledge related to quantum computing?
- Do you think the knowledge about quantum computing might change in the future? Explain why or why not.

6. Gravitational Waves: How Much Science Is Too Much?

Back in 1915, Einstein published his groundbreaking paper on general relativity. The theory states that massive objects can bend the way things around them move, affecting spacetime. When Einstein was exploring the mathematical implications of this theory, he discovered that there should be entities described as "gravitational waves," which are waves caused by those massive objects rippling through the fabric of spacetime. Flash forward 100 years into the future, when the Laser Interferometer Gravitational-Wave Observatory (LIGO) made the first direct observation of gravitational waves brought on by two black holes colliding (Chu, 2016). The discovery of gravitational waves represents a monumental achievement in humanity's understanding of the cosmos, offering unprecedented insights into elusive celestial phenomena. However, the ensuing discussions surrounding gravitational wave research have led to a debate regarding the allocation of resources and the societal value of further exploration in this domain.

Advocates for continued inquiry into gravitational waves emphasize the unparalleled scientific opportunities afforded by this breakthrough. They contend that the detection of gravitational waves opens new vistas for studying phenomena such as black hole mergers and neutron star collisions—events that were previously undetectable. Moreover, proponents argue that investing in gravitational wave research aligns with broader social values, fostering global interconnectedness and stimulating curiosity in STEM (Science, Technology, Engineering, and Mathematics) fields, thereby nurturing scientific literacy and innovation on a global scale.

This is in opposition to those who believe that those purposes are not unimportant compared to other matters. They question the allocation of resources to a field they perceive as tangential to more pressing societal concerns. They argue that resources would be better directed towards addressing urgent issues such as healthcare, environmental conservation, and poverty alleviation. Moreover, skepticism has been cast on the credibility of the detection methods employed, particularly concerning LIGO. Dr. Jackson, an esteemed astroparticle and gravitational physicist, has raised doubts about the authenticity of the detected gravitational waves, characterizing them as a mere "illusion" (Brooks, 2018).

- If you were in a position to make a decision, would you continue funding gravitational wave research efforts?

Yes No

- Why or why not? Provide arguments.
- Professor Ponso disagrees with you. What would he say to try and convince you otherwise?
- How would you reply?
- Do you think your decision might change in the future? Why or why not? Explain.
- Why did scientists reach different conclusions even though they were looking at the same data?

7. Should We Reach for the Stars?

The burgeoning industry of space tourism holds promise for opening up access to space for private individuals, unlocking the door for individual space exploration. However, the rapid growth of space tourism also raises questions about its environmental impact and the distribution of resources. As companies like SpaceX and Virgin Galactic race to commercialize space travel, there have emerged two opposing groups.

Supporters of space tourism tout its potential to drive technological innovation, stimulate economic growth, and propel scientific advancement. They argue that the commercialization of space travel will spur the development of cost-effective technologies, such as reusable rockets and spaceports, thereby lowering the barriers to entry for space exploration. Moreover, proponents contend that space tourism holds immense educational value, inspiring future generations of scientists and engineers to pursue careers in STEM fields. Additionally, they envision space tourism as a catalyst for international cooperation and collaboration, fostering diplomatic relations and accelerating scientific discoveries on a global scale (Shah, 2023).

Conversely, detractors of space tourism raise concerns about its environmental impact and resource allocation. Critics highlight the significant carbon emissions and pollutants generated by rocket launches, exacerbating environmental degradation, and contributing to climate change. Moreover, the proliferation of space tourism vehicles adds to the growing problem of orbital debris, posing hazards to existing satellites and spacecraft in orbit and increasing the risk of collisions (Pultarova, 2021). This group argues that addressing pressing terrestrial challenges, such as climate change and environmental sustainability, should take precedence over the commercialization of space travel.

You are an American citizen. The government is considering a nationwide referendum to determine whether to provide public funding for space tourism initiatives. The outcome of this vote will decide whether federal resources will be allocated to support the growth of this industry or if the government will prioritize environmental and climate initiatives instead.

- Do you vote with or against funding?
- Why or why not? Provide arguments.
- Professor Ponso disagrees with you. What would he say to try and convince you otherwise?
- How would you reply in defense of your stance?
- Do you think your decision might change in the future? Why or why not? Explain.
- Why did scientists reach different conclusions even though they were looking at the same data?

8. Particle Therapy

Particle therapy is a promising frontier in the realm of cancer treatment. Using carbon ions or protons, it involves the use of particle accelerators to precisely target tumors with high radiation, leaving the surrounding healthy tissue intact (Mohan & Grosshans, 2017). But not everyone sees particle therapy as a panacea. Opinions are split between those who cite its health benefits and others who decry its cost.

Advocates of particle therapy extol its precision and efficacy, particularly in cases where critical organs are involved. By utilizing carbon ions or protons, particle therapy minimizes collateral damage to adjacent tissues, leading to fewer severe side effects compared to traditional treatments like chemotherapy. Furthermore, proponents emphasize its potential to reduce the risk of secondary cancers, making it particularly well-suited for treating pediatric cancers where long-term effects are of paramount concern (Thomas & Timmermann, 2020). The emphasis on quality-of-life positions particle therapy as a superior addition to the arsenal of cancer treatment modalities.

Conversely, detractors raise concerns about the logistical and financial implications of widespread adoption of particle therapy. The substantial investment required for particle accelerators, ongoing maintenance, and staff training poses a significant financial burden, potentially straining healthcare systems and exacerbating healthcare disparities. Critics argue that prioritizing expensive therapies like particle therapy for a select group of patients may divert resources away from preventive measures, early detection initiatives, and treatments with proven cost-effectiveness. Additionally, skeptics question the lack of robust evidence demonstrating particle therapy's superiority over conventional cancer treatments, highlighting the need for further research to validate its efficacy and long-term outcomes (Ofuya et al., 2019).

You are a doctor and have been contacted by a particle therapy modality salesperson. You bring that up in your daily meeting with the hospital director.

- Do you try and persuade the hospital director to adopt particle therapy as the main form of cancer treatment?
- Why or why not? Provide arguments.
- Dr. Downer disagrees with you. What would she say to try and convince you otherwise?
- How would you reply?
- How can you explain that scientists reached different conclusions even though scientists were all looking at the same data about the effects of particle therapy?
- How do scientists know about the effects of particle therapy?
- How certain are scientists about the knowledge related to particle therapy?
- Do you think the knowledge about particle therapy might change in the future? Explain why or why not.

References

Brooks, M. (2018, October 31). *Exclusive: Grave doubts over LIGO's discovery of gravitational waves.* NewScientist. Retrieved from https://www.newscientist.com/article/mg24032022-600-exclusive-grave-doubts-over-ligos-discovery-of-gravitational-waves/

Brownell, V. (2019, March 11). *Quantum computing could change the way the world uses energy.* Quartz. Retrieved from https://qz.com/1566061/quantum-computing-will-change-the-way-the-world-uses-energy

Burgan, C. (2023, November 16). *Red hat: Final NIST quantum guidelines coming in 2024.* MeriTalk. Retrieved from https://www.meritalk.com/articles/red-hat-final-nist-quantum-guidelines-coming-in-2024/#:~:text=Presman%20explained%20that%20while%20it's,start%20preparing%20now%2C%20he%20said

Chu, J. (2016, February 11). *Scientists make first direct detection of gravitational waves*. MIT News. Retrieved from https://news.mit.edu/2016/ligo-first-detection-gravitational-waves-0211

Davies, N. (2023, July 19). *Here's what quantum computing is—And how it's going to impact the future of work, according to a software engineer.* Fast Company. Retrieved from https://www.fastcompany.com/90925065/heres-what-quantum-computing-is-and-how-its-going-to-impact-the-future-of-work-according-to-a-software-engineer

Elhegazy, H., & Kamal, M. (2021, October 12). Implementing nuclear power plants (NPPs): State of the art, challenges, and opportunities. *Innovative Infrastructure Solutions, 7*(1). https://doi.org/10.1007/s41062-021-00611-z

EnergySage. (2021). *What are the environmental benefits of solar energy—And what are the drawbacks?* Retrieved from https://www.energysage.com/solar/health-environmental-benefits-of-solar-energy/

Gordon, D. (2023, June 3). *Solar panels - An eco-disaster waiting to happen?* BBC News. Retrieved from https://www.bbc.com/news/science-environment-65602519

Hill, J. S. (2023, December 11). *Solar costs are now nearly 30 per cent lower than cheapest fossil fuel option, EY says.* Renew Economy. Retrieved from https://reneweconomy.com.au/solar-costs-are-now-nearly-30-per-cent-lower-than-cheapest-fossil-fuel-option-ey-says/

Jackson, S. (2023, April 13). *Solar panel pros and cons: Find out if they're worth it for you*. CNET. Retrieved from https://www.cnet.com/home/energy-and-utilities/solar-panel-pros-and-cons-find-out-if-theyre-worth-it-for-you/

Leman, M. (2022, March 18). *6 reasons why nuclear energy is not the way to a green and peaceful world*. Greenpeace. Retrieved from https://www.greenpeace.org/international/story/52758/reasons-why-nuclear-energy-not-way-green-and-peaceful-world/

Mohan, R., & Grosshans, D. (2017). Proton therapy – Present and future. *Advanced Drug Delivery Reviews, 109*, 26–44. https://doi.org/10.1016/j.addr.2016.11.006

Office of Nuclear Energy. (2022, October 3). *5 fast facts about spent nuclear fuel*. Energy.gov. Retrieved from https://www.energy.gov/ne/articles/5-fast-facts-about-spent-nuclear-fuel#:~:text=Spent%20nuclear%20fuel%20can%20be,%2C%20such%20as%20France%2C%20do

Ofuya, M., McParland, L., Murray, L., Brown, S., Sebag-Montefiore, D., & Hall, E. (2019). Systematic review of methodology used in clinical studies evaluating the benefits of proton beam therapy. *Clinical and Translational Radiation Oncology, 19*, 17–26. Retrieved from https://www.ncbi.nlm.nih.gov/pmc/articles/PMC6660607/

Pultarova, T. (2021, July 26). *The rise of space tourism could affect Earth's climate in unforeseen ways, scientists worry*. Space.com. Retrieved from https://www.space.com/environmental-impact-space-tourism-flights

Rhodes, R., & Beller, D. (2000). The need for nuclear power. *Foreign Affairs, 79*(1), 30–44. https://doi.org/10.2307/20049612

Shah, R. (2023, November). *Space tourism market outlook from (2024 and 2034)*. Future Market Insights. Retrieved from https://www.futuremarketinsights.com/reports/space-tourism-market

Thomas, H., & Timmermann, B. (2020). Paediatric proton therapy. *The British Journal of Radiology, 93*(1107), 20190601. https://doi.org/10.1259/bjr.20190601

U.S. National Security Agency. (2023, August 21). *Post-quantum cryptography: CISA, NIST, and NSA recommend how to prepare now*. National Security Agency/Central Security Service. Retrieved from https://www.nsa.gov/Press-Room/Press-Releases-Statements/Press-Release-View/Article/3498776/post-quantum-cryptography-cisa-nist-and-nsa-recommend-how-to-prepare-now/

Union of Concerned Scientists. (2013, March 5). *Environmental impacts of solar power*. Retrieved from https://www.ucsusa.org/resources/environmental-impacts-solar-power

Chapter 5
Environmental Science

Abstract This chapter investigates numerous environmental issues, with a focus on conservation, water resources, and the overall environmental effect of human activities. Topics including deforestation, climate change, light pollution, waste management, and urbanization emphasize the fragile balance between economic expansion and environmental sustainability. The conversation also covers Lebanon's environmental issues, such as the Naameh landfill catastrophe, open-burning habits, and the debate over electric bikes and scooters as pollution remedies. Water-related concerns such as overfishing, shortages, and wastewater reclamation highlight the importance of sustainable resource management. This chapter also dives into technology solutions such as geoengineering and renewable energy transitions, as well as the economic and societal trade-offs associated with environmental regulations. This chapter promotes critical thinking about the future of environmental protection and sustainable development by offering several points of view.

Keywords Environmental conservation · Climate change · Deforestation · Sustainable development · Water scarcity

This chapter delves into socioscientific scenarios in environmental science, focusing on critical concerns such as conservation, water resources, management, and environmental effect. Under the subject of environmental conservation, issues are investigated such as deforestation, reforestation efficacy, soil pollution, climate change, light pollution, and waste management, with an emphasis on Lebanon's garbage dilemma and waste disposal procedures.

Moving on to water resources, overfishing and industrial fishing are discussed, while water management scenarios include topics such as water scarcity, desalination, wastewater reclamation, water filtration, and environmental concerns surrounding projects such as Lebanon's Bisri Dam. The chapter finishes by presenting larger environmental repercussions, including the dual nature of pesticides, acid rain, Cool Seal, Cloud Seeding, and the environmental consequences of quarrying. Each scenario encourages examination of the socioscientific elements, stressing the interconnectedness of environmental concerns and their societal consequences.

R. Khishfe, *The Complexities of Socioscientific Issues in Education*, Springer Texts in Education, https://doi.org/10.1007/978-3-031-88006-3_5

A. **Environmental conservation**

1. Deforestation
2. Deforestation (Effectiveness of reforestation)
3. Climate Change
4. Light Pollution
5. Waste Incineration
6. The Bird Hunting Dilemma: Striking a Balance between Conservation and Tradition in Lebanon
7. The Garbage Crisis
8. Incineration
9. Waste Disposal
10. Electric Bikes and Scooters: A Controversial Solution to Air Pollution in Lebanon
11. Naameh Landfill (Waste management)
12. Waste Management in Lebanon
13. Overpopulation
14. Garbage Open-Burning in Lebanon
15. Urbanization
16. Economic Development and the Sea
17. Geoengineering (Technological Solution to Climate Change)
18. Renewable Energy Transition (Balancing Economic Growth and Environmental Sustainability)
19. Beach Privatization in Lebanon

B. **Water resources**

20. Overfishing
21. Industrial Fishing

C. **Water management**

22. Water Scarcity and Desalination
23. Water Scarcity and Access to Clean Water
24. Wastewater Reclamation
25. Water Filtration
26. Water Scarcity in Lebanon
27. Water Pollution
28. Flood management and Resilient Infrastructure
29. Water Fluoridation

D. **Environmental Impact**

30. The Dual Nature of Pesticides
31. Acid Rain
32. Cool Seal
33. Cloud Seeding
34. Quarrying

35. Global Warming

A. Environmental conservation

1. **Deforestation**

Deforestation is the process of purposefully clearing forested lands to cater to more domestic needs. In 2022 alone, thousands of square kilometers were deforested in the Amazon rainforest, which is known as the world's lungs (Worldwide Fund for Nature-Brazil, 2022).

Groups in support of deforestation are mainly farmers who view the deforestation of tropical rainforests such as the Amazon rainforest in Brazil as an opportunity to expand cattle ranches and soybean farms (Food and Agriculture Organization, 2023a). They believe that with an increasing rate of population growth they are in need to increase production and contribute to the national economy. Deforestation can be used to clear land for farming and cattle grazing, which can help people in need of money by giving them a source of food and income. Research has indicated a relationship in certain areas between the rates of avoiding deforestation and poverty (Andersen, 2002). Deforestation may be seen as the sole way to get firewood for cooking and heating in areas where access to alternative energy sources is restricted. Research highlights how important it is to handle fuelwood sustainably and have access to renewable energy sources (Bailis et al., 2005). This group denies the environmental aspects by justifying them through advanced business models or accusing scientists of falsifying information or inflating the risks associated with deforestation.

On the other hand, groups against deforestation see the environmental catastrophe that can follow it. Firstly, deforestation threatens biodiversity worldwide, as it puts many kinds of plants and animals in danger of going extinct. For example, ten percent of all known species on Earth are found in tropical rainforests, such as the Amazon rainforest, but similar risks are occurring in many other regions where deforestation is taking place (Laurance et al., 2022). On a global level, forests—particularly the Amazon rainforest—absorb carbon dioxide, a major greenhouse gas. Deforestation releases this carbon into the atmosphere, contributing to global warming. Also, deforestation affects water supplies and might result in more flooding in certain areas and droughts in others due to climate change. Economically, numerous nations depend on the income from tourism that their forests produce. Deforestation damages the scenic appeal and biodiversity of forests, which negatively affects local economies and the tourism sector. Furthermore, deforestation has devastating effects on indigenous communities, as many rely on forests for their livelihoods, traditions, and way of life. The destruction of their ancestral lands disrupts their ability to hunt, fish, and gather food, threatening their cultural heritage and forcing many to face displacement.

Your government plans on deforesting an area and requires citizens to vote on the issue.

- Would you vote "yes" to deforesting the area?

YES NO

- Explain and justify your position on deforestation.
- An environmentalist or investor disagrees with your stance on deforestation. How will they persuade you with their explanation of their position on deforestation?
- How would you respond and communicate with them to express your disagreement with their position?
- Do you think your decision might change in the future? Why or why not? Explain.
- Why did scientists reach different conclusions even though they were looking at the same data?

2. **Deforestation (Effectiveness of Reforestation)**

Deforestation stands out as a major contributor to alarming levels of climate change and global warming. Scientists highlight the drastic reduction in forested land, now covering only 30 percent, which is not sustainable given our current practices. If we keep cutting down forests the way we are right now, experts predict that we will lose most of the world's forests in about 540 years (United Nations Food and Agriculture Organization, 2005). The annual destruction of millions of acres of forest land, predominantly impacting tropical forests, has significantly reduced the amount of land that once constituted 15 percent of Earth's total area to only 6 or 7 percent. This environmental degradation poses two significant threats: the release of carbon dioxide into the atmosphere and the ongoing loss of biodiversity. According to the World Economic Forum, deforestation is a substantial source of carbon emissions that accelerates climate change."

A proposed solution involves restoring forests through conservation and widespread tree planting, known as reforestation. Advocates argue that efforts to stop tree cutting and plant trees in areas that were previously deforested could help reduce carbon emissions by stabilizing atmospheric levels. A 2019 study found that reforestation could contribute to carbon storage by drawing CO_2 out of the atmosphere as forests mature. This group believes that reforestation represents a promising complement to technological efforts to combat climate change.

However, some researchers and climate scientists question the ability of reforestation alone to reverse the effects of climate change or achieve zero-net emissions. They highlight that the estimates from studies may overstate the potential for tree planting due to issues like the availability of suitable land, much of which is already occupied by other vegetation. Additionally, climate scientists emphasize that planting trees on a large scale could have unintended consequences. For instance, trees can absorb heat due to their dark surfaces, exacerbating warming effects in regions like snow-covered areas near the poles. This suggests that while tree planting is a helpful tool, it should not be viewed as a single solution to the climate crisis.

- As a scientist, do you agree that reforestation is an effective way of fighting deforestation and restoring our climate?

YES NO

- Explain and justify your position.
- Another scientist has disagreed with your decision, in what way would you explain your position and convince him of your decision?
- How do you think he would respond in a way that may convince you?
- Suppose you agree with the scientists who opposed reforestation as the most effective way to reduce carbon emissions, what alternatives can you propose?
- Do you think, if provided more data, you would change your decision? Justify.
- Is there anything else you would want to know about this issue that might help you decide, or change your decision?

3. **Climate Change**

Earth's surface temperatures have risen by more than 1 °C in recent years, accompanied by a significant increase in atmospheric levels of greenhouse gases, including carbon dioxide (CO_2) and methane (CH_4). Two contrasting perspectives exist regarding the cause of this surge in greenhouse gases and its impact on climate change.

One viewpoint attributes the rise to human activities, particularly the consumption of non-renewable energy sources like oil and coal. The unique isotopic ratio of CO_2 from human activities, as confirmed by the Intergovernmental Panel on Climate Change (IPCC), distinguishes it from naturally occurring CO_2. The IPCC and the US Environmental Protection Agency emphasize that almost all the increased greenhouse gases over the last 150 years are a result of human activities.

However, an opposing perspective contends that human-produced CO_2 is absorbed by the Earth's carbon sinks, such as oceans and forests, mitigating climate effects. Research from the *Asia-Pacific Journal of Atmospheric Sciences* challenges climate models that overestimated the impact of CO_2 emissions on warming. It suggests that approximately half of human-caused CO_2 emissions have already been reabsorbed by the Earth's carbon sinks. From 2002 to 2011, 26% of human-induced CO_2 emissions were specifically absorbed by the world's oceans, leading to ocean acidification, which threatens marine ecosystems by causing the loss of corals and other species.

Additionally, a study published in the *Proceedings of the National Academy of Sciences* indicates that elevated CO_2 levels are fostering increased forest growth, potentially lowering atmospheric CO_2. However, scientists caution that this negative feedback effect may not continue indefinitely. Studies suggest that as CO_2 levels rise further, other limiting factors, such as nutrient availability and water stress, could slow forest growth, reducing their capacity to absorb CO_2 (*New Scientist*, 2023). This aligns with the perspective that asserts climate change is not solely driven by human activity but involves complex interactions with the Earth's natural mechanisms.

- Do you think that relying on natural processes, such as oceans and forests absorbing CO_2, is enough to counter the effects of climate change?
- Explain and justify your position.
- Professor Ponso disagrees with your viewpoint. What would he say to try and convince you otherwise?
- What would you reply to fortify your argument?
- How can you explain that scientists reached different conclusions even though scientists were all looking at the same data about the effects of climate change?
- How do scientists know about the effects of climate change?
- How certain are scientists about the knowledge related to climate change?
- Do you think the knowledge about climate change might change in the future? Explain why or why not.

4. **Light Pollution**

Light pollution, an often-overlooked issue with significant consequences for both current and future generations, is responsible for the deaths of millions of birds each year. Controversial opinions surround the topic of light pollution, with some considering it an emerging environmental problem that will present serious challenges in the future.

One group views the issue as tremendously impactful. The disruptions caused by light pollution pose threats to wildlife, impacting the physiology of animals and plants whose life rhythms are influenced by the alternation of light and dark. The rapid increase in light pollution over the years has deprived people of the experience of marveling at dark skies, a phenomenon that scientists define as the presence of artificial or "anthropogenic" light in the night environment. This not only disrupts ecosystems but also obscures the starlight in the night sky, leading to a loss of connection with cultural heritage and the potential to inspire future generations.

On the contrary, some individuals argue that the severity of light pollution is overstated and that there is insufficient and inconsistent information to fully understand its long-term impact. They may believe that the current research is flawed or incomplete, with data that could be outdated, misinterpreted, or lacking in context. Critics also point to the challenges of determining the true extent of light pollution's effects due to the complexity of ecological systems and varying regional circumstances. As a result, they maintain that more rigorous and up-to-date scientific research is necessary before implementing major policy changes or large-scale mitigation efforts.

- Do you believe that addressing light pollution should be a priority in environmental conservation efforts?

Yes No

- What reasons do you have for believing so?
- Professor Ponso disagrees with your position. What would he say to make you change your opinion?
- What would you say to convince him of your belief?
- How can you explain that scientists reached different conclusions even though scientists were all looking at the same data about the effects of light pollution?
- How do scientists know about the effects of light pollution?
- How certain are scientists about the knowledge related to light pollution?
- Do you think the knowledge about light pollution might change in the future? Explain why or why not.

5. **Waste Incineration**

The world generates a staggering 1.3 billion tons of garbage annually, with 60% of this waste finding its way into landfills. However, this method is deemed insufficient and alternative solutions, such as incineration, are being considered. Incineration, a waste treatment method involving the burning of organic materials in waste, releases ashes, flue gas, and heat. The ashes, primarily composed of inorganic components, are treated as solid particulates, and carried by flue gas. Before release into the air, the flue gas undergoes a cleaning process to remove particulates and gaseous contaminants. The released heat can be harnessed for burning additional waste or generating electricity.

Supporters of waste incineration, particularly environmental scientists, argue its environmental friendliness citing several reasons. They emphasize that incineration reduces waste volume by 80%, efficiently manages waste in any weather, occupies minimal space, generates electricity, diminishes pollution, controls odor and noise, eliminates harmful germs and toxic clinical chemicals, prevents methane gas production, reduces global warming, eliminates the need for waste transportation, diminishes groundwater contamination, effectively collects and recycles metals, produces less than 10% residual ash with various applications, and can be monitored and filtered to comply with emission limits.

In contrast, opponents of waste incineration highlight its drawbacks. They argue that it is an expensive method and raises environmental concerns due to the production of greenhouse gases such as carbon dioxide, acid gases, heavy metals, nitrogen oxides, carcinogenic dioxins, and particulates. Detrimental effects on public health, including cancer, birth defects, reproductive dysfunction, and neurological problems, are also cited. The production of ash waste containing harmful substances, like heavy metals, is considered problematic, and opponents suggest that waste incineration discourages recycling by requiring large volumes of waste to sustain the process.

The decision to adopt waste incineration for garbage disposal in a village is a contentious issue. While some support the method for its efficiency, reduced waste volume, and energy generation, others oppose it due to environmental and health concerns, as well as its potential to discourage recycling. The ultimate stance on the matter depends on a careful consideration of the advantages and disadvantages, as well as the specific circumstances and priorities of the village.

- Do you approve of your city's plans to incinerate the garbage in your village?

Yes No

- Explain and justify your decision.
- What would be the argument of another scientist, who disagrees with your position?
- What would you reply to the scientist to explain that your decision is right?
- Do you think that the knowledge about the effects of waste incineration might change in the future? Why or why not?
- Would you change your decision in the future?

6. **The Bird Hunting Dilemma: Striking a Balance between Conservation and Tradition in Lebanon**

Lebanon boasts a longstanding cultural affinity with the animal kingdom, where birds play a crucial role in the country's rich biodiversity. That has made the country a prominent hunting hotspot in the region. Presently, the Lebanese government is discussing whether to pass legislation to greatly reduce bird hunting. This has sparked a division among the public, with two distinct groups voicing their perspectives.

The first group supports this legislation, seeing as Lebanon boasts many different species and seeing those numbers dwindle would negatively impact our perception of our heritage, our environment, and our tourism sector. Lebanon is reputed to host around migrating bird species and is also a critical location for many millions of migratory European birds. Due to hunting, many species have been greatly affected. Furthermore, birds do not exist in isolation, but are also pollinators. Decreasing bird populations could lead to deforestation and desertification. That would lead to a trickle-down effect on ecosystems, affecting other animals in addition to plants.

On the other end, detractors disagree. For starters, they claim that these regulations are ineffective. Scientific research from the University of Kent in the United Kingdom has shown that in such cases, hunters merely take their practice elsewhere. On top of that, hunting is not just a pastime, but several families in Lebanon relies on to survive. Moreover, this group believes that controlled hunting could lead to economic growth by generating conservation revenues. For instance, it could be applied as an activity to raise conservation revenues. Finally, hunting is part of Lebanese culture and continues to be practiced by people of different socioeconomic and educational backgrounds, who view it as part of their identity.

As a resident of Lebanon, where the government mandates residents to vote for or against this issue, the outcome will determine whether the proposed legislation will be implemented within the next three years.

- Would you vote to implement stricter regulation in Lebanon?

YES NO

- Explain and justify your decision.
- Professor Mary disagrees with your position. How could she explain her position to illustrate the reasons supporting it and convince you?
- What would you reply to Professor Mary to explain your position is right?
- Why do you think there are different positions from scientists who are studying the same information about bird hunting?
- Do you think knowledge about bird hunting might change in the future? Please clarify.
- Do you think you might change your decision in the future? Please clarify.

7. **The Garbage Crisis**

The world produces over two billion tons of solid waste every year, enough to fill over 800,000 Olympic sized swimming pools. Around 2 million tons are from the total waste in Lebanon (United Nations Development Programme, 2020). The garbage crisis is a serious issue the world is facing. It has a huge impact on the environment and health of humankind at large.

Different governments have attempted to tackle this problem in different ways. Some have elected to use garbage to generate electricity. Others have taken to performing of recycling.

A group of Lebanese people thinks that garbage can be more than a liability—that we can make something valuable out of it. This group supports the burning of garbage to help generate electricity. Currently, Lebanon spends over $420 million annually to deal with solid waste (Sofian, 2023). This amount does not provide any immediate financial returns. This group mainly provides burning as the solution, but there are other, cleaner ways. For instance, incineration can be relegated only to organic materials, in a process known as pyrolysis. Another way of breaking down organic material is thermal depolymerization. These two processes are less pollutant than simply burning everything in landfills, and more efficient than waiting for the garbage to decompose and making use of the methane exhausted as fuel.

The other group is comprised of scientists who are against transforming garbage into electricity as they see that it has many side effects on the country and the citizens. Burning garbage without removing inorganic materials beforehand can send extremely noxious chemicals into the air, such as sulfur dioxide and heavy metals. Additionally, some disposed items might contain hazardous substances like furans, which can cause skin toxicity and endocrine disruption. Finally, this group sees this strategy as wasteful, because it takes the focus away from recycling and more sustainable waste management practices.

Suppose you live in Lebanon, where the government is considering using waste-to-energy technology. You are asked to vote on this issue.

- Would you support the use of waste-to-energy technology?

Yes No

- Explain and justify your decision.
- Another researcher opposes your viewpoint. How could he describe his stance and persuade you by demonstrating the grounds for it?
- What would you say to the scientist to convince him of your position?
- How can you explain why experts came to various findings on the effects of employing the waste-to-energy process even though they were all looking at the same data?
- Do you think our understanding of the waste-to-energy process will evolve in the future? Please clarify.
- Do you believe you will reconsider your decision in the future? Please elaborate.

8. **Incineration**

Waste incineration, also known as Waste to Energy, involves the burning of solid waste.

Lebanon faces a significant challenge with garbage disposal, prompting various eco-friendly non-governmental organizations to act—but there are detractors.

Critics of waste-to-energy incinerators argue that this method is ineffective for Lebanon. The combustion of waste releases cancerous gases like dioxins, exacerbating the country's already high cancer rates. Furthermore, Lebanon's predominantly organic waste, constituting around 70% of the total, cannot be incinerated due to its moisture content. This includes high-calorific plastics and cardboard, leading to the loss of recyclable materials and missed job opportunities. Concerns about health issues and the difficulty in tracking pollution further discourage the installation of incinerators, as expressed by the Water Management Coalition (WMC) in Lebanon.

On the other hand, supporters of incinerators argue that they can mitigate water pollution, reduce carbon footprints, and play a role in decreasing land contamination. They emphasize the presence of filters to trap pollutants, disputing claims of environmental harm. Additionally, incinerators' ability to operate in various weather conditions is highlighted, offering an advantage during Lebanon's rainy winters. Despite these debates, some NGOs propose alternative solutions, such as waste sorting at the source, promoting the 3 Rs (reduce, reuse, recycle), and biologically treating organic waste to minimize overall waste.

Suppose you live in Lebanon. The government is considering building incinerators and wants its citizens to vote on the issue.

- Are you with or against building incinerators in Lebanon?
- Explain and justify your decision.
- Another scientist, Professor Saliba, disagrees with your decision. How could she explain her position to illustrate the reasons supporting it and convince you?
- What do you reply to Professor Saliba to explain that your decision is right?
- Do you think the knowledge on incinerators will change in the future?
- Do you think you might change your decision in the future?

9. **Waste Disposal**

Recycling is the process of collecting sorted waste and processing it instead of discarding it randomly, involving the creation of new products from used materials. This method offers significant advantages to both the community and the environment.

A substantial group supports recycling for various reasons. Primarily, recycling plays a crucial role in reducing pollution by decreasing the volume of waste disposed of in landfills and incinerators. Studies indicate that industries can achieve a 73% reduction in air pollution and a 35% decrease in water pollution through waste recycling. Additionally, recycling contributes to preserving natural resources and saving energy. Neglecting to recycle aluminum cans and newspapers results in the wasteful use of energy equivalent to that produced by around 15 power plants. Furthermore, some recyclables, such as aluminum and glass, can be recycled multiple times while maintaining the same efficiency as when used for the first time. Ultimately, recycling generates economic benefits by creating job opportunities for unemployed individuals in waste collection, processing, manufacturing, and the purchase of new products.

On the contrary, those who oppose waste recycling argue that there is no landfill crisis to begin with. They contend that existing landfills and incinerators are not harmful to nature and the environment. Additionally, opponents highlight the high costs associated with building the necessary equipment and tools for the recycling process. Studies suggest that recycling plastic may emit toxic chemicals into the air, contributing to air pollution, rendering recycling an inefficient solution to address air pollution. Moreover, detractors argue that recycling imposes a burden on families, citing a lack of time or knowledge to sort and recycle waste. Some complain that there are no avenues to sell or reuse recyclables, deeming the investment of time and money in recycling unnecessary.

In your country, the government is pondering what to do regarding waste and wants its citizens to weigh in on the issue.

- Do you agree with waste recycling?

Yes No

- What is your position about whether these processes and plans of waste disposal are good or bad? Do the positives outweigh the negatives, or vice versa?
- Another researcher concerned with this issue disagrees with you. How would he/she develop his/her arguments to convince you of his/her position?
- What would your response be to this researcher who has an opposing position with yours?
- Why do you think there is a difference in scientists' positions about such issues?
- Do you think that knowledge about recycling might change in the future? Why or why not?

10. **Electric Bikes and Scooters: A Controversial Solution to Air Pollution in Lebanon**

Recently, there has been a noticeable increase in the use of electric bikes and scooters as modes of transportation in Lebanon. This surge can be attributed to the escalating prices of gas and transportation services like taxis. Individuals have found these methods to be more cost-effective and time-efficient (due to traffic). The air quality in Lebanon is exceptionally poor, with its toxicity, according to atmospheric chemist Najat Saliba, being linked to the fact that breast cancer cases are diagnosed in Lebanese women, on average, ten years younger than the global average. Approximately 50% of Lebanon's air pollution is caused by cars, with older vehicles contributing more to pollution than those with newer technology. The average age of cars in Lebanon is 19 years old. Despite the natural defense mechanisms of our respiratory system, prolonged exposure to air pollutants can compromise these mechanisms, leading to health issues such as asthma, bronchitis, and lung cancer. According to the U.S. Environmental Protection Agency, annual deaths related to air pollution range from 150,000 to 350,000 people. It has been revealed that pollutant levels inside cars in traffic can be up to 18 times higher than outside. Other countries have addressed this issue by imposing high taxes on old cars or mandating a transition to newer models after a certain vehicle age. However, such measures are not feasible for many people during an economic collapse, coupled with a lack of public transportation options. Consequently, the reliance on old cars for transportation remains significant. Thus, a debate surrounds the promotion of this trend as a solution to Lebanon's air pollution problem.

Reducing the use of these cars would significantly enhance air quality, as observed during lockdowns when people had to refrain from using their cars. This is why some people view electric scooters and bikes optimistically as contributors to improved air quality. Plus, at a manageable cost, bikes and scooters can be charged through solar panels. Bikes and scooters could help out a sizeable portion of the population, as a significant amount of car trips is taken by individual people, such as university students, and can be reduced.

However, concerns arise about the safety of individuals using them, given that the conditions of roads in Lebanon are not ideal for such modes of transportation. Additionally, some argue that they cannot be considered entirely green options since they require electricity for charging, and in Lebanon, electricity is predominantly produced through diesel generators, a particularly toxic method that results in a high concentration of NO_2, classified as a criteria pollutant detrimental to human health. Additionally, arguments persist that bikes and scooters are inconvenient for individuals with children, disabilities, or those traveling in large groups.

Suppose you live in Lebanon. You see an advertisement encouraging people to use bikes.

- Would you be willing to give up the convenience of having/using a car?

Yes No

- Explain and justify your position.
- If someone were to argue against it, what would their supporting reasons be?
- Is there evidence that we can account for some of this person's concerns and still advocate for giving up cars?
- With the elections coming up, do you prioritize environmental legislation when you look at political campaigns? What argument would you give to someone that says there are more urgent issues to care about than air pollution?
- Do you think your decision might change in the future? Why or why not? Explain.
- Why did scientists reach different conclusions even though they were looking at the same data?

11. **Naameh Landfill (Waste management)**

The origins of the current waste management crisis in Lebanon date back to 1994 when the Lebanese Council for Development and Reconstruction hired Sukleen, a private garbage management firm, to collect waste in Beirut. Sukleen was awarded a $3.6 million contract, almost double the amount municipalities might have paid for garbage collection in their districts. In 1996, the Lebanese government granted approval for Sukleen to create a short-term dump near Naameh to store trash from Beirut and Mount Lebanon until a more sustainable solution could be found. Naameh received approximately 3000 tons of solid trash daily, more than three times its intended capacity. Despite locals' objections, the landfill's closing date was repeatedly postponed until July 17, 2015, due to political gridlock preventing the government from finding a permanent garbage management plan. Residents disagreed with the government's decision to use Naameh as a landfill.

On one hand, scientists argue that the trash could be used to generate electricity, reducing greenhouse gases. The government claimed there was no alternative place to store garbage, and Naameh was a short-term solution for the crisis. Additionally, it was a cost-effective way to maintain cleanliness in the city, considering Lebanon's limited large spaces for garbage storage.

On the other hand, residents had a different perspective. They began experiencing environmental and health issues, including stomach aches, dizziness, vomiting, coughing, and allergies, dubbing the landfill as "the landfill of death." Cancer became a prevalent cause of death in the village. Locals, particularly the less fortunate, felt abandoned, and the poor village faced a significant health crisis. Concerns about smell, leachate, gas emissions, groundwater pollution, odor releases, and poisonous chemicals highlighted the environmental hazards. Unmanaged trash started burning, releasing methane gas and causing an increase in fires, threatening the health of the region's population. The noise produced by trucks became bothersome, and the vehicles violated rules by crossing through the town, emitting an unpleasant organic smell. Protesters forcibly closed the landfill, contributing to Lebanon's ongoing rubbish crisis.

Suppose you are a Lebanese citizen. Authorities are considering re-opening the Naameh landfill to store garbage.

- Would you vote for the re-opening of the landfill?

Yes No

- Explain and justify your position.
- Another resident disagrees with you; how could they explain their position to illustrate the reasons supporting it and convincing you?
- What would you reply to the citizen to explain that your position is right?
- Do you think that the knowledge on this issue (long-term effects on the environment and humans' health) might change in the future?
- Do you think you might change your decision in the future? Please explain.
- What do you think is the ideal solution to solve the garbage issue in Lebanon? Do you think it is possible to implement it?

12. **Waste Management in Lebanon**

Waste management in Lebanon presents a multifaceted challenge with severe environmental and public health implications. As the BBC reports, the global scale of municipal solid waste production is staggering, with over 2 billion tons generated annually, equivalent to filling over 800,000 Olympic-sized swimming pools (McGrath, 2019). Lebanon is not exempt from this issue, facing a situation where more than 37% of materials in the waste stream are recyclable, yet only a portion of less than 10% is recycled (Abbas et al., 2019). Instead, a significant portion ends up in overflowing landfills or is burnt, permeating the air with hazardous compounds. Lebanon's struggle with waste management intensified notably in 2015, marked by the closure of the Naameh landfill without a viable backup plan. Consequently, garbage piled up on streets, forming infamous "garbage mountains" across the country. Most people can agree that waste can lead to harmful outcomes, such as an increase in cancer cases (Germanos & Azzi, 2024), but the contested issue is the question of how to deal with the waste. Mainly, there is a team that is against landfills, and another which sees them as a viable solution.

Those against landfills cite their environmental and public health-associated impacts. For starters, landfills, which receive 48% of collected waste, are closely linked to soil and water pollution, posing significant challenges for Lebanon's ecosystems. Studies have shown the presence of toxins like dioxin and dibenzanthracene in the atmosphere due to landfill usage, leading to chronic diseases and cancer risks (Abbas et al., 2019). The foul odor emitted by landfills, as witnessed before the closure of the "Nehmeh" landfill, has sparked public outrage due to its adverse effects on residents' health. Moreover, the conducive environment for bacterial breeding in landfills raises concerns about potential outbreaks of new epidemics or pandemics. The presence of landfills also increases the risk of fire hazards in fertile land (Massoud et al., 2016). On the positive side, forgoing landfills in favor of recycling optimizes land utilization, as exemplified by the transformation of former landfill sites into green spaces like the Al-Saudi garden in Saida (Naharnet Newsdesk, 2014). As for the question of cost, recycling offers significant economic benefits, with the potential for annual income ranging from $65 million to $272 million. Additionally, embracing proper waste management practices like recycling has the potential to add 1900–2500 new jobs over nine years, aiding Lebanon economically (Khalil, 2022). Plus, recycling can conserve resources, which leads to a reduction in energy consumption and minimizes the need for new raw material extraction.

On the other end, those that tolerate the status quo of relying on landfills highlight concerns regarding the country's ability to afford advanced waste management technologies. The economic crisis has strained resources, making it challenging to invest in necessary equipment and tools. Public health specialists have conducted research in developing countries, which showed how prioritizing waste management will divert funding away from other pressing socioeconomic issues such as healthcare or education (Ferronato & Torretta, 2019). With Lebanon generating 7000 tons of waste daily (Sbeih, 2023), the financial burden of proper waste management would heavily affect the government, as the cost is estimated at approximately $120 USD per ton (Abbas et al., 2019), further compounds the issue. Moreover, the political complexity of the country, rooted in sectarianism, hampers reform efforts, hindering the implementation of effective waste management laws (Bodetti, 2019). This team also argues against recycling by citing the lack of public awareness and participation, and how costly it would be to bring everyone up to par.

Suppose you are a Lebanese citizen. The Lebanese government has just announced that it will be extending some of its landfills to face the ongoing garbage crisis, instead of finding new methods to manage the nation's waste.

- Do you support or oppose the government's decision?
- Explain and justify your decision.
- Dr. Saliba disagrees with your position. How could you explain her position to illustrate the reasons supporting it and convince you?
- How do you think your position will affect the environment in the future?
- Do you think your decision might change in the future? Why or why not? Explain.
- Why did scientists reach different conclusions even though they were looking at the same data?

13. **Overpopulation**

Many recent debates have revolved around the question of whether overpopulation is an issue or not, leading to the categorization of overpopulation as an environmental concern requiring attention. Overpopulation occurs when the population expands exponentially without regard for the limits of the existing environment or community.

Given the controversial nature of the overpopulation topic, two distinct perspectives have emerged. The first perspective is supported by a group of scientists who argue that overpopulation can have certain economic and social benefits. They argue that sustained overpopulation has a positive impact on the environment. According to these scientists, there is a direct proportional connection between population growth and economic status. They contend that higher population density contributes to elevated economic status and improved innovation. Sectors like agriculture also benefit from overpopulation, as they experience expansion. In Lebanon, the agriculture sector constitutes 5% of GDP, employing approximately 20% of the population. Moreover, these scientists assert that urbanization is a consequence of overpopulation, leading to the development of cities. In Lebanon, the population increases at a rate of 1% per year.

In contrast, another perspective vehemently believes that overpopulation is a critical issue requiring serious attention. This group argues that overpopulation places considerable strain on urban infrastructure, with challenges such as increased demand for natural resources, higher rates of solid waste generation, and pressure on water supplies becoming more evident in densely populated areas. They state that an average of 1.57 million tons of waste is generated per year, with 65% originating in urban areas. Furthermore, the population increase in Lebanon leads to higher water demands, with two-thirds of natural resources experiencing bacterial contamination, especially within urban areas. Additionally, the influx of displaced refugees in Lebanon is exacerbated by overpopulation and economic challenges. Currently, 1.3 million Syrian refugees in Lebanon struggle in extreme poverty, with nine out of ten Syrian refugee families living in impoverished conditions.

You live in Lebanon. You are watching a talk show where people are discussing overpopulation.

- After reviewing the various perspectives on the scenario, do you endorse the idea of overpopulation?
- Explain and justify your position.
- If someone holds a contrasting opinion to yours, what points might they raise to support their argument?
- How would you counter the opposing viewpoint?
- Is there any possibility that you might reconsider your stance on increasing the human population? Provide an explanation for your response.

14. **Garbage Open-Burning in Lebanon**

Open burning involves burning unwanted objects in the open air, emitting smoke and dangerous gases directly into the atmosphere, causing environmental damage.

This practice, often used for incinerating household waste as a disposal method, releases uncontrolled and untreated combustion into the atmosphere, posing dangers to the environment and human health, especially with materials like plastics containing harmful substances.

In 2015, when garbage overwhelmed Beirut's streets, Lebanon's waste problem gained international attention. Locally, opinions were divided on whether burning rubbish is a problem or a solution.

Supporters argue that Lebanon must continue burning waste to prevent a dangerous buildup, stating that a 2020 Human Rights Watch report on the trash crisis warned that Beirut's major landfills were nearing capacity. Additionally, this group sees burning trash as a potential solution to Lebanon's electrical crisis. They cite Denmark's clean waste-to-energy incinerators as an example, suggesting that Lebanon could produce energy and fuel from waste with sufficient funds, addressing the long-standing electricity shortage.

Opponents argue that burning garbage has severe environmental and public health consequences. It is said Human Rights Watch found in 2017 that burning waste at over 150 open dumps risked nearby residents' health, causing issues like chronic obstructive pulmonary disease, coughing, throat irritation, skin conditions, and asthma. The opponents stress the adverse effects on human health, citing the ingestion of mercury by fish, a toxic substance released during open burning, leading to various diseases affecting the brain, kidney, heart, and lungs. Furthermore, the opposition points out the global repercussions, such as large-scale soot generation contributing to global warming by reducing the environment's reflecting nature. They argue against the notion that rubbish can be used for clean energy, emphasizing that waste-generated energy is still contaminated, according to the Conversation Law Foundation, which states that incineration relies on a constant flow of waste, including items that could be composted or recycled, and uses a substantial amount of energy to operate.

Imagine you live in Lebanon and are thus affected by this issue. In your opinion, is garbage open burning a problem or a solution? Explain.

- If you were to suggest a solution or alternative for open burning in Lebanon, what would you say? Support your answer with evidence.
- Some scientists claim that open burning is a great source for using waste in producing electricity and fuel or energy in general, do you agree with that?
  - If yes, would you think the government can take some measures to make this happen in Lebanon? Elaborate.
  - If not, why? Justify your reasoning.
- Despite the passage of a solid waste management law in 2018 banning the open burning of waste in Lebanon, municipalities still engage in the practice, posing health risks to nearby residents, especially children and older persons. Suggest other effective ways to prevent the violating institution from committing such harmful acts.

- How would you reply to the idea about waste piling up of the group supporting the burning of garbage? Give an example and illustrate.
- Do you think your decision might change in the future? Why or why not? Explain.
- Why did scientists reach different conclusions even though they were looking at the same data?

15. **Urbanization**

With rising education levels, climate change, and an increased focus on service jobs, urbanization and overpopulation in cities have become prominent topics of debate since the 2000s, with over 4 billion people worldwide living in urban areas (Ritchie et al., 2024). To put that into perspective, urban areas occupy only 1% of the Earth's total land area (Ritchie, 2018). While urbanization can offer individuals better economic opportunities, healthcare access, and improved quality of life, it has also raised concerns about public health, environmental degradation, and rising living costs. The rapid acceleration of urbanization has thus led to complex and often conflicting perspectives regarding its overall effects on social, environmental, and economic conditions.

The group in support of urbanization has explicitly examined its positive impacts regarding the health, economic, and educational aspects. When it comes to health, a study conducted around cervical cancer screening among women in Lebanon revealed that women living in metropolitan areas like Greater Beirut reported a 59.7% of women that used pap smears for screening as compared to those living in rural areas, which tallied 39.1% (Bou-Orm et al., 2018). This indeed shows how people residing in urban cities have more access to healthcare services than rural area dwellers, enabling them to prevent serious illnesses. It's also worth noting that from an economic perspective, a report published by UN-Habitat on Lebanon's urbanization shows that rural-urban migration poses positive impacts on the economy in terms of the increase in trade and capital movements (UN-Habitat, 2011). Additionally, having skilled people in proximity drives innovation and competition, further strengthening the economy (think Silicon Valley and Wall Street). Moreover, the level of education one can attain is different in the city. A research study published by the Lebanese American University [a university in Lebanon] has shown that families are allowing their children to move into urban areas, particularly Beirut, to enroll in universities and pursue their degrees there (Youssef, 2019). Furthermore, because of education being the main indicator of qualification in the city instead of lineage and inherited assets, there is increased social mobility.

However, there are detractors. This group perceives urbanization in Lebanon as a threatening phenomenon, posing various destructive effects on the country itself and its population. Opponents have argued that research studies have revealed an association between urbanization and alarming health repercussions. From a public health aspect, The Lebanese Center for Policy Studies has indicated that the rapid urbanization in Lebanon has increased the population's density in the urban areas contributing to the development of overcrowded cities and dwellers with poor living conditions (Fawaz, 2017). This in return has increased the levels of stressors as well as pollution-related diseases among individuals (Fawaz, 2017). Moreover, from an environmental aspect, Annahar News report that specific areas in Lebanon, particularly Zouk, Antelias and Hazmieh, have lost their green spaces as they encountered residential and commercial development (Khamis, 2018). In addition, from a financial aspect, urbanization has led to an increase in the population density in specific areas, thus contributing to an increase in rent prices (Dagher, 2020). In Beirut, rent prices are reported to be four times the minimum wage of an individual, making them unable to afford a well-established residence and prompting them to live in poor urban slums (Dagher, 2020), which comes with its own slew of social problems (like drug-related crimes, gang activity, and fragile infrastructure that could lead to health issues and even casualties).

Imagine you are a Beirut native, and the government is discussing the possible mandate of a law that would curb urbanization. Your friend is from rural Bekaa and is always telling you about how they would love to move to the capital. The government puts the issue to a vote.

- Do you vote for or against urbanization?
- Provide evidence.
- Professor Lee disagrees with you. What does she say to convince you of her position?
- How do you defend your position?
- How can you explain that scientists reached different conclusions even though scientists were all looking at the same data about the effects of urbanization?
- How do scientists know about the effects of urbanization?
- How certain are scientists about the knowledge related to urbanization?
- Do you think the knowledge about urbanization might change in the future? Explain why or why not.

16. **Economic Development and the Sea**

The delicate balance between economic development and environmental conservation in coastal regions has become a contentious issue, particularly considering escalating environmental threats such as ocean acidification (Hu, 2022). On one hand, proponents of economic development argue for the importance of leveraging coastal resources to stimulate growth, stimulate the job market, and improve living standards, such as in cities like Jounieh, Lebanon. On the other hand, opponents raise concerns about the potential negative impacts of unchecked development on fragile marine ecosystems and biodiversity. This debate underscores the complex trade-offs and competing priorities inherent in managing coastal areas and underscores the need for holistic approaches that reconcile economic imperatives with environmental sustainability.

Advocates for economic development in coastal regions emphasize the significant socioeconomic benefits that can accrue from harnessing coastal resources for tourism, fisheries, and infrastructure development. They argue that responsible development practices can stimulate economic growth, alleviate poverty, and improve the quality of life for coastal communities. Additionally, they contend that economic development can enhance resilience and adaptive capacity in the face of environmental threats, such as climate change and natural disasters, by providing resources for mitigation and adaptation measures.

The opposing group says that those ventures mainly benefit already-powerful groups, such as resort chains and the government. They raise concerns about the disproportionate impacts of coastal development on vulnerable communities, including indigenous peoples, small-scale fishers, and marginalized populations, who often bear the brunt of environmental degradation and loss of livelihoods-like in Bali (Warren & Steenbergen, 2021) and other renowned destinations. Moreover, those detractors also highlight the inherent risks and negative externalities associated with exploiting coastal resources for short-term gains. They argue that unsustainable development practices, such as coastal construction, deforestation, and pollution, can degrade marine ecosystems, exacerbate environmental degradation, and undermine long-term ecological sustainability.

You live in a coastal town, where the mayor is considering allowing a resort chain to set up shop along the shore. The mayor says that this will drive tourism and lead to more jobs, but thirty fishermen's houses will have to be torn down. He puts the issue to a vote.

- Do you vote for or against the mayor's plans?
- Provide evidence.
- Patricia disagrees with you. What does she say to convince you of her position?
- How do you defend your position?
- Do you think your decision might change in the future? Why or why not? Explain.
- Why did scientists reach different conclusions even though they were looking at the same data?

17. **Geoengineering (Technological Solution to Climate Change)**

In the ongoing debate over climate change, geoengineering has arisen as a contentious technical answer. While some support geoengineering as a way of mitigating the consequences of human climate change, others are cautious, citing a variety of issues.

Proponents claim that geoengineering is a viable way to combat climate change, particularly in view of mounting evidence of human influence on the Earth's climate. They see it as an important complement to existing methods such as emission reduction and adaptation initiatives. Furthermore, they emphasize the potential benefits of geoengineering, such as solar radiation management (SRM) and carbon dioxide removal (CDR), in tackling climate-related issues.

However, skeptics have legitimate worries about the hazards connected with geoengineering activities. They warn against fast deployment, citing the uncertainty and potential consequences of such large-scale technology interventions. They also emphasize the significance of contemplating alternatives to climate change, such as limiting fossil fuel usage and investing in renewable energy sources. Critics say that geoengineering should not be prioritized above attempts to address the core causes of climate change, and they push for further study and international governance to monitor the development and deployment of geoengineering technology. As the discussion continues, stakeholders must address the complex social, political, and ethical components of geoengineering to make educated decisions and engage the public on this crucial topic.

- Are you with or against using geoengineering as a way to tackle climate change?
- Explain and justify your position.
- Some people disagree with your position. How could they justify their position to convince you and illustrate the reasons supporting their position?
- What would you reply to these people to explain that your position is right?
- How can you explain that scientists reached different conclusions even though scientists were all looking at the same data about the effects of geoengineering?

- How do scientists know about the effects of geoengineering?
- How certain are scientists about the knowledge related to geoengineering?
- Do you think the knowledge about geoengineering might change in the future? Explain why or why not.

18. **Renewable Energy Transition (Balancing Economic Growth and Environmental Sustainability)**

The transition to renewable energy sources is a complex and multifaceted endeavor that involves navigating competing priorities, particularly balancing economic growth and environmental sustainability. Proponents of a rapid shift toward renewable energy emphasize its role in mitigating climate change, reducing air pollution, and creating new 'green' jobs that could stimulate long-term economic development. Conversely, critics argue that the existing fossil fuel-based energy infrastructure remains more reliable and affordable, pointing to the challenges of renewable energy's variability and the costs of large-scale investment. These contrasting views underscore an ongoing societal debate, with each side emphasizing different aspects—economic security, technological feasibility, environmental necessity, and equity-making this transition far from straightforward.

Those that support transitioning to renewable energy wholly and expeditiously state that it is crucial for addressing climate change and ensuring long-term sustainability. Proponents also highlight the potential for new, "green," jobs, stimulating economic growth, and improving public health by reducing air pollution. Empirical investigations of the trilemma energy balance (sustainability, affordability, and security) and renewable energy transitions provide detailed insights. They show that actions encouraging the three aforementioned criteria boost economic development while improving environmental results (Khan et al., 2022). Notably, nations such as Switzerland, Sweden, and Denmark have shown leadership in developing sustainable energy regulations, creating a precedence for others to follow.

On the other hand, detractors of immediate transitioning to renewable energy argue that the current energy infrastructure, primarily based on fossil fuels, is more reliable and cost-effective. They point out the challenges and high costs associated with developing and scaling up renewable energy technologies. Concerns include the variability of renewable sources like wind and solar, which can lead to energy supply instability. This group also emphasizes the economic impact, particularly on industries and communities dependent on fossil fuel production. They advocate for a gradual transition, focusing on improving energy efficiency and integrating renewables within the existing energy mix to avoid economic disruptions and ensure energy security.

- Are you in favor of putting economic development ahead of environmental sustainability in the near term? Or do you feel that long-term environmental protection should take precedence, even if it results in slower initial economic progress?
- Explain and support your stance.
- Some people disagree with your position. How could they justify their position to convince you and illustrate the reasons supporting their position?

- What would you reply to these people to explain that your position is right?
- How can you explain that scientists reached different conclusions even though scientists were all looking at the same data about the effects of renewable energy transition?
- How do scientists know about the effects of renewable energy transition?
- How certain are scientists about the knowledge related to renewable energy transition?
- Do you think the knowledge about renewable energy transition might change in the future? Explain why or why not.

19. **Beach Privatization in Lebanon**

Beach privatization refers to the process of limiting free access to a coastal area by a private entity, such as a resort, hotel, or restaurant. Privatization has long been a highly debated topic in the Lebanese scene, as it is not only a morally controversial issue but a legal and environmental one.

Advocates for the privatization of beaches state that it is beneficial to the economy, especially through tourism. Further, it is noted that pollution on private beaches is minimal as they are cleaned regularly. A study of beaches in Lebanon showed that 9 of the 13 unsafe beaches, which were unsafe because of their amount of fecal matter, were public (Keuchkerian, 2022). Additionally, beach privatization is alleged to generate governmental revenues through lease agreements, taxes, and licensing fees.

As for the other side, those against privatization state that beaches are public property by law, and that through privatization, access to the beach is restricted to many citizens. "Privatization makes it almost impossible for residents to access the beaches since it costs between $15 to $20 to access one of these private beaches," says Paola Rebeiz, a longtime activist with NAHNOO (Courtat et al., 2023). In fact, there have been many protests regarding the clear disregard for the law and the public. The above argument also presents the assumption of a false dichotomy, in which a clean beach must be private, negatively impacting the drive towards environmentally friendly communities, and limiting environmental education and awareness opportunities. Public beaches can be as clean as private ones through public incentives and NGOs. Furthermore, this group argues that beach privatization is one of the many faces of corruption in Lebanon. Five million square meters are perpetrated by members of all sects, with only a few of them paying a mere $156,000 annually to the Lebanese government (Hourany, 2023).

You are a Lebanese citizen. A group of resorts want to set up branches in Beirut. That would line the coast of Beirut with beaches that have been privatized, restricting access to public beaches.

- What do you say to that?
- Explain and justify your decision.
- Amelia disagrees with your position. How would she explain her position to illustrate the reasons associated with it?
- How would you respond to Amelia?
- Do you think your decision might change in the future? Why or why not? Explain.
- Why did scientists reach different conclusions even though they were looking at the same data?

B. Water resources

20. **Overfishing**

Overfishing, acknowledged as a pressing global issue, involves the excessive extraction of fish from oceans, rivers, and other bodies of water, surpassing sustainable levels required for natural replenishment. This unsustainable practice carries significant and far-reaching consequences for the environment. To elaborate, overfishing disrupts marine ecosystems, leading to the depletion of fish populations and the disruption of the food chain. Additionally, it poses a threat to biodiversity as it often targets large and commercially valuable species, consequently affecting the overall balance of marine life. Beyond environmental impacts, overfishing can result in economic and social ramifications, significantly impacting the livelihoods of fishing communities worldwide. Among those effects is decreased availability of fish, especially in the case of vulnerable communities. In addition to that, there is the exploitation of those disadvantaged communities by fishing corporations.

To avoid those consequences, a growing body of activists has been advocating for the close monitoring of fishing. Currently, 7.5% of the ocean is comprised of Marine Protected Areas (MPAs) (United Nations Food and Agriculture Organization, 2023b). MPAs are designated areas of the ocean (or lakes/estuaries) where fishing and other human activities are heavily restricted. This is seen as one of the most effective ways to replenish the ocean and curb overfishing. The second method experts suggest is banning fishing company subsidies. Subsidies are sub-companies that are part of a larger network. In the fishing industry, they flood the market with fishing gear and vessels, encouraging overfishing. Finally, this group implores governments to establish traceability standards. These are related to documentation and licenses obtained from authorities to ensure that no unreported or unregulated fishing takes place.

But there is a group of detractors who find that those measures will do more harm than good. For starters, they posit that due to the growing global population, demand for fish is greater than ever. After all, one billion people suffer from vitamin D deficiency (Institute of Medicine, 2011), which fish can alleviate, as they are an important source of the "sun vitamin." On top of that, tackling overfishing through banning subsidies would affect revenues and job security. Finally, strict traceability standards would make things much harder for small-time fishermen that merely fish to put food on the table and make a meager income.

Your government is thinking of imposing measures against overfishing, and it requires its citizens to vote on the issue.

- Do you support the measures taken against overfishing?

Yes No

- What compelling arguments would you put forth to defend your stance, considering its diverse impacts on the environment, economy, and society?
- Dr. Nancy disagrees with you. What could she say to try and persuade you of her stance?
- How would you reply in defense of your original viewpoint?
- How do scientists know about the effects of overfishing?
- Do you think your decision might change in the future? Why or why not? Explain.
- Why did scientists reach different conclusions even though they were looking at the same data?

21. **Industrial Fishing**

Fishing and its impact on both humans and the environment have been the subject of debate among scientists worldwide. While fishing provides crucial nutritional benefits-offering essential nutrients like omega-3 fatty acids and supporting human health-it also poses environmental risks, such as the decline of aquatic species and ecosystem degradation. As a result, there is an ongoing discussion about how to balance the benefits of fishing for human health with its environmental consequences, with many scientists emphasizing sustainable fishing as a way to address both challenges.

Scientists claim that fishing is essential because it makes one of the most nutrient-dense superfoods readily available to the public. Fish is a high-quality protein that is rich in omega-3 fatty acids, vitamin D, vitamin B2, calcium, phosphorus, and many minerals like iron, zinc, magnesium, potassium, and iodine. It is also a good source of iron, zinc, and vitamin B2. Fish consumption can reduce blood pressure and the risk of heart attacks and strokes, offering significant health benefits. Fish consumption is linked to critical nutrients that support and preserve brain and heart health, which is crucial given that approximately 108 million people in the United States alone have hypertension. Since our bodies cannot produce omega-3 fatty acids, we must obtain them from our food. Additionally, fish may be able to help with other medical issues, including goiter.

Another group of scientists contends that fishing is actively reducing the number of cod, tuna, and other species at an alarming rate, in addition to endangering the environment. Scientists have identified fishing, particularly on an industrial scale, as a major factor in climate change due to the decline of aquatic species and the destruction of their habitats. More marine life equals more carbon sequestration, which is important since it stores carbon emissions. The earth's greatest carbon reservoir, the ocean, holds about 38,000 billion tons of carbon. The temperature of the oceans is rising due to carbon emissions in the atmosphere, which affects aquatic organisms and, when combined with overfishing, threatens ocean life, according to scientists. The symbiotic relationship between coral and aquatic life must also be noted; when fish are scarce, corals suffer because fish play a vital role in maintaining the balance of their ecosystems. For instance, herbivorous fish consume algae that would otherwise grow unchecked on coral reefs. If these algae are allowed to overtake corals due to the reduction of fish populations, the corals become smothered and lose their ability to thrive. Conversely, healthy coral reefs support marine life by providing habitats and sustaining fish populations, creating a feedback loop. This mutual dependence highlights how overfishing can lead to a loss of fish and coral diversity, contributing to the collapse of marine ecosystems. This leads to a decline in ocean life and, according to scientists, makes it impossible for us to undo the harm we have caused. Depletion of fish also raises the ethical question of society's failure to safeguard and preserve the environment and its value. Vulnerable communities are forced to struggle because their natural food supply is jeopardized. The pollution of the food supply reduces the safety of the food and seriously jeopardizes the health of people who rely solely on natural fishing.

- As a scientist, do you believe the current scale of fishing should continue as it is, be reduced, or be managed differently to balance human health and environmental sustainability?

Yes No

- Explain and justify your position.

- Another scientist disagrees with your decision. What points could this scientist make to convince you to side with them?
- What is your response as to why your decision is correct?
- Do you think your decision might change? Why or why not? Explain.
- Why did scientists reach different conclusions?

C. Water management

22. **Water scarcity and Desalination**

In a coastal city experiencing water scarcity, there is a growing debate about implementing large-scale desalination to meet the population's water demands.

The group in support of desalination, representing water security experts and public health officials, argues that desalination is a viable solution to address water scarcity. They point to studies that demonstrate the reliability and resilience of desalinated water supply, particularly in arid and water-stressed regions. This group argues that desalination can help meet the growing population's increasing water demands and ensure a safe and secure water supply, reducing the risk of waterborne diseases and enhancing public health.

On the other hand, the group representing environmental advocates argues against desalination due to potential environmental impacts. They cite studies that highlight the negative consequences of desalination on marine ecosystems, including harm to marine organisms through the intake and discharge of highly saline brine and the release of chemicals used in the desalination process. To elaborate, after desalination, the remaining concentrated salt is usually pumped back again into the sea. If there are fish close to the site of salt re-entry, there is a chance they could be harmed or killed. This group emphasizes the need to prioritize ecosystem conservation and sustainable water management strategies that reduce water demand and enhance water efficiency to mitigate water scarcity.

- Would you support the implementation of large-scale desalination as a solution to address water scarcity in your region?

Yes No

- Explain and justify your decision.
- Professor Ponso disagrees with your decision. What would he say to make a case for himself?
- How would you defend your original viewpoint in response to his position?
- What factors do you think should be considered when deciding about implementing large-scale desalination in a coastal city?
- What alternatives would you propose to address water scarcity, and why?
- How do scientists know about the effects of water scarcity and desalination?
- Do you think your decision might change in the future? Why or why not? Explain.
- Why did scientists reach different conclusions even though they were looking at the same data?

23. **Water Scarcity and Access to Clean Water**

Water scarcity and inaccessibility to clean water are major environmental issues that are often overlooked. This is alarming, as access to clean water is essential for preventing life-threatening diseases such as cholera and syphilis. While our planet is often referred to as the "water planet," the reality is that clean and freshwater accounts for only about three percent of the water on Earth, according to the U.S. Geological Survey (National Geographic Education, n.d.). Furthermore, the rate at which we are depleting these limited freshwater supplies is accelerating due to growing populations, industrialization, climate change, and unsustainable water management practices.

The situation is worsened by examples like the *Flint Water Crisis* in the United States, which underscores how failures in infrastructure can lead to devastating consequences for public health when access to clean drinking water is compromised. Despite being a fundamental resource, access to clean and safe water remains a challenge for millions of people worldwide.

Two clashing perspectives on this issue exist: economic prosperity versus environmental sustainability.

Economic Perspective

Industrial and agricultural activities are major contributors to global water usage. According to the World Bank Organization, irrigated agriculture accounts for about 70% of global freshwater withdrawals, making it a key driver of water scarcity in many regions. Proponents of this economic perspective argue that strict water conservation measures would hinder industries reliant on high water use, such as agriculture, manufacturing, and tourism. They emphasize that such measures could lead to job losses and negatively impact economic growth.

Environmental Sustainability Perspective

Conversely, advocates for environmental sustainability prioritize environmental and human health by promoting water conservation, reducing pollution, and implementing innovative water-saving technologies. This group argues that minimizing water waste and adopting sustainable practices can prevent long-term consequences such as widespread water scarcity, ecological damage, and declining water quality.

Both perspectives highlight the need to balance economic activity with sustainable water use in order to address global water challenges.

- Do you believe that reducing water usage will eventually lead to economic deterioration and job losses in various economic industries, including agriculture?

Yes No

- Please explain and justify your position.
- If you have chosen to advocate for environmental sustainability, what recommendations would you propose to address concerns about the economic impacts of water conservation on industries reliant on heavy water use?
- Do you think your decision might change in the future? Why or why not? Explain.
- Why did scientists reach different conclusions even though they were looking at the same data?

24. **Wastewater Reclamation**

Wastewater reclamation serves as an alternative method for providing drinkable water in countries facing water resource depletion due to factors such as climate change and pollution. In these situations, the treatment of wastewater, which includes domestic and commercial water usage, can transform it into safe water for agricultural uses, for instance.

Proponents of using wastewater for industry and agriculture argue that it is important to consider places where no other viable water options are available. As for the question of public acceptance, awareness campaigns done in Namibia have shown that people were generally accepting of using it. Plus, leaving wastewater contaminated, even if not directly consumed, can increase the risk of infections and poisonings, impacting global health.

Advocates of wastewater treatment explain that untreated wastewater might contaminate other water sources, degrading soil and crop quality, and endangering the health of other animals. They also note that the overall cost of treating wastewater is deemed lower than the potential costs associated with environmental degradation.

One major concern revolves around the potential toxicity of industrial wastewater. Critics argue that while advanced treatment processes can significantly reduce the levels of industrial toxins in water, eliminating them entirely remains a significant challenge, particularly for highly persistent contaminants. Additionally, untreated wastewater, which often contains pathogens such as bacteria and viruses, poses serious health risks and can contribute to diseases like diarrhea and cholera if not properly addressed.

To address concerns about the quality of treated water, a monitoring system with expert oversight is implemented. However, skepticism exists in your country regarding the establishment of an independent committee due to sociopolitical structures that may enable corruption. The lack of trust between the population and the government's operating skills further complicates the implementation of effective wastewater treatment.

- Would you be willing to use water that comes from a wastewater treatment plant? In what ways (i.e., drinking, cleaning, agriculture, etc.)? Would you use it for one purpose and not the other?
- Explain and justify your position.
- If someone were to argue against the use of such water, what would their supporting reasons be?
- Can these reasons be countered by other evidence?
- How do sociopolitical factors influence such decisions? Can an argument apply to all countries?
- How can you explain that scientists reached different conclusions even though scientists were all looking at the same data about the effects of wastewater reclamation?
- How do scientists know about the effects of wastewater reclamation?
- How certain are scientists about the knowledge related to wastewater reclamation?
- Do you think the knowledge about wastewater reclamation might change in the future? Explain why or why not.

25. **Water Filtration**

Water filtration involves the removal or reduction of particulate matter, including suspended particles, parasites, bacteria, algae, viruses, and fungi, as well as other unwanted chemical and biological pollutants from contaminated water, producing safe and clean water for a specified purpose. Water filtration is used in a variety of contexts, including sewage treatment plants and industrial applications, but recently there has been a push to implement filtration systems at the household level to make tap water drinkable.

The group advocating for filtered tap water believes that, amid economic crises, it could help save money on bottled water and reduce the production of water bottles. Discarding plastic bottles instead of recycling them contributes to pollution, and the associated energy and production costs increase the carbon footprint. Having access to safe drinking water at home is a significant advantage. Furthermore, some filtration systems are certified to eliminate hazardous contaminants found in well water, providing safe and clean water.

On the other hand, opponents argue that domestic water filtration systems require consistent cleaning and filter replacement. Without proper maintenance, dirty or clogged filters can lose their effectiveness, potentially allowing pollutants, including heavy metals, to pass through into the water supply. Negligent filter care may also result in mold formation when filters become moist and warm. A study reported significantly higher biofilm growth and bacterial colony counts in poorly maintained filtration systems-up to 10,000 times higher than in untreated tap water. While these findings raise concerns about potential health risks, such contamination does not always directly result in severe illness.

- Would you vote for water filtration implementation?
- Explain and justify your decision.
- Another scientist, Professor Pafero, disagrees with your position. How could he explain his position to illustrate the reasons supporting it and convince you?
- What would you reply to Professor Pafero to explain that your position is right?
- How can you explain that scientists reached different conclusions even though scientists were all looking at the same data about the effects of water filtration?
- How do scientists know about the effects of water filtration?
- How certain are scientists about the knowledge related to water filtration?
- Do you think the knowledge about water filtration might change in the future? Explain why or why not.

26. **Water Scarcity in Lebanon**

Lebanon has been referred to as the "Middle East's water tower." Despite an annual average rainfall exceeding 800 million cubic meters (MCM), the country faces an increasing vulnerability to water shortages, especially during dry summers. This is attributed to a lack of water storage infrastructure, rising demand, and the growing impacts of climate change. Water scarcity has become a prominent national concern. Lebanon currently ranks 149th globally in terms of water availability, a troubling position given the significant rainfall the country receives annually.

The group supporting urgent action on water shortages argues that it is a major challenge affecting millions of people. UNICEF Lebanon warns that over four million residents are at risk of losing access to clean water. According to Shaban (2019), the crisis stems from an undefined hydrological cycle and insufficient hydrological data, which make it difficult to develop effective strategies and policies. This group suggests that solving Lebanon's water problem requires implementing innovative measures and adopting a comprehensive national water policy that considers all hydrological factors, including sustainable water resource management.

On the other hand, opponents believe that addressing water supply issues relies more on natural factors such as consistent rainfall and the existence of extensive river basins. They argue that constructing large-scale infrastructure, such as dams, while potentially increasing water storage capacity, comes with significant environmental risks and high costs. Opponents also claim that measures currently in place are sufficient to manage the issue and prevent severe water scarcity. Some suggest that investing in water treatment and optimizing existing water systems might be more practical than implementing entirely new water supply methods.

You live in Lebanon, and your city plans to implement a new water supply strategy, requiring residents to vote for or against it:

- Would you support the proposed water supply strategy in your region?

Yes No

- Explain and justify your position.
- If you meet a professor who does not agree with your point of view, how could they justify their position by illustrating their reasoning?
- What evidence will you use to show the professor that your position is valid?
- Do you believe that people's perceptions of water supply strategies will change in the future?
- Is it possible for your decision to change in the future?

27. Marine Pollution: The Crisis of Chemical and Trash Contamination

Marine pollution is a global issue affecting oceans worldwide, primarily due to chemical contamination and trash. These two types of pollution pose significant threats to human health, the environment, and the economy. Human activities, such as fertilizer usage on farms, contribute to chemical runoff into waterways, eventually reaching the ocean. Lebanon, like many nations, faces serious pollution challenges. Improperly managed garbage contributes to contaminated waterways, and a

2020 study by the National Council for Scientific Research revealed that eight out of 31 coastal sites in Lebanon are extremely polluted.

One group opposes imposing stricter regulations on factories. They argue that water treatment is expensive and time-consuming, requiring new equipment and specialized personnel, which could reduce production rates and harm manufacturers' profitability. This would, in turn, hinder economic growth and potentially lead to job losses. For industries, it is often cheaper to discharge untreated waste into water sources than to process or treat it properly.

On the other hand, the opposing group emphasizes that pollution's long-term consequences outweigh the immediate economic benefits. Coastal cities like Beirut, Tripoli, and Tyre are heavily affected by factory waste, leading to marine ecosystem imbalances, such as some species adapting to polluted environments while others decline. This imbalance threatens biodiversity and food security. Additionally, consuming fish from polluted waters poses health risks, such as food poisoning, and contributes to rising greenhouse gas emissions in the Mediterranean Sea.

- Would you vote in favor of this motion or against? Would you vote in favor of banning factories from discharging untreated water into open water sources?

Yes No

- Explain the reasoning behind your answer.
- The mayor disagrees with your decision. What would he say to convince you of his position?
- How would you defend your original stance?
- How do scientists know about the effects of water pollution?
- Do you think your decision might change in the future? Why or why not? Explain.
- Why did scientists reach different conclusions even though they were looking at the same data?

28. **Flood management and Resilient Infrastructure**

In the face of escalating climate change impacts and increasing urbanization several solutions have emerged, such as flood management strategies, which include warning systems and awareness campaigns designed to reduce the impact of floods on communities (examples include sensors, communication systems, and emergency plans). In addition to flood management strategies, resilient infrastructure provides another barrier against incurring tremendous flood-related losses. Examples of resilient infrastructure include flood barriers, elevated buildings, and flood-proof doors in homes. As extreme weather events become more frequent and severe, the need to develop robust flood management strategies and resilient infrastructure systems has never been more urgent. But as with most solutions, there are disadvantages one has to bear in mind.

Advocates for flood management and resilient infrastructure emphasize the importance of proactive planning and investment to reduce vulnerability and enhance the adaptive capacity of communities facing flood risks. They argue that strategic flood management measures like seasonal rounds of flood preparedness training among citizens can help mitigate flood impacts, protect lives and property, and promote sustainable development. Moreover, they contend that resilient infrastructure, such as raised electrical systems, robust drainage systems, and levees (which are slanted embankments built on bodies of water) can minimize disruption and economic losses during extreme weather events, ultimately saving lives and taxpayer dollars in the long run.

Alternatively, critics of flood management and resilient infrastructure raise concerns about the feasibility, cost, and unintended consequences of implementing such measures. They argue that large-scale infrastructure projects, such as levees, dams, and floodwalls, may have adverse environmental and social impacts, including habitat destruction, displacement of communities, and exacerbation of flood risks downstream (Ma et al., 2022). Additionally, they raise questions about the equity and affordability of resilient infrastructure, particularly for low-income and marginalized communities who may lack the financial resources to implement adaptive measures, such as elevating buildings or installing flood-proof doors.

Your town is part of a regional consultation process regarding the construction of a dam that will cut through a forest hosting endangered native species. The regional government is seeking public input through a vote.

- Would you vote in favor of or against this plan?
- Explain the reasoning behind your answer.
- Jake disagrees with your decision. What would he say to convince you of his position?
- How would you defend your original stance?
- Do you think your decision might change in the future? Why or why not? Explain.
- Why did scientists reach different conclusions even though they were looking at the same data?

29. **Water Fluoridation**[1]

The fluoridation of water involves adding fluoride to public drinking water. This issue is controversial and has been the cause for many court cases.

The group in favor of water fluoridation considers fluoridation as a safe and inexpensive way to prevent tooth decay for all citizens during their lifetime. They point out that many distinguished national and international scientific organizations support fluoridation. Further, this group argues that scientific research shows that water fluoridation reduces tooth decay and cavities and prevents dental disease (Aoun et al., 2018).

The group against fluoridation considers it unethical because it is a form of involuntary medication. They also point out that fluoridation does not have Food and Drug Administration (FDA) approval. Further, this group argues that scientific

[1] This scenario was first published in Khishfe, R. (2012). Relationship between nature of science understandings and argumentation skills: A role for counterargument and contextual factors. *Journal of Research in Science Teaching, 49*(4), 489–514. doi:https://doi.org/10.1002/tea.21012

research shows harmful effects of fluoridation, such as possible links to cancer (Bassin et al., 2006). Further, fluoride can be toxic to pregnant women and is linked to lower IQ in newborns (Till & Green, 2021). Furthermore, it adversely affects thyroid and reproductive functions. In addition, adding fluoride to drinking water makes it impossible to know how much fluoride a person takes. At the same time, the effects of fluoride are topical so it would be better to apply fluoride directly to the teeth using the toothpaste instead of ingesting it through water. It is also claimed that scientists who release papers showing the harmful effects of water fluoridation are being silenced by industry scientists.

Your city plans on adding fluoride to drinking water and requires residents to vote for or against this issue. If they get enough votes, then water fluoridation will be effective for the next five years.

- As a scientist, would you vote for adding Fluoride to drinking water in your city?

YES NO

- Explain and justify your decision.
- Another scientist, Professor Ponso, disagrees with your decision. How could he explain his position to illustrate the reasons supporting it and convince you?
- What would you reply to Professor Ponso to explain that your decision is right?
- How can you explain that scientists reached different conclusions even though scientists were all looking at the same data about the effects of water fluoridation?
- How do scientists know about the effects of water fluoridation?
- How certain are scientists about the knowledge related to water fluoridation?
- Do you think the knowledge about water fluoridation might change in the future? Explain why or why not.
- As a scientist, do you think you might change your decision in the future? Explain why or why not.

D. Environmental Impact

30. **The Dual Nature of Pesticides**

Pesticides are chemical compounds designed to eliminate pests, which include insects, rodents, fungi, and unwanted plants (weeds). Many different pesticides are utilized globally, serving various purposes. Pesticides find applications in public health to combat disease vectors like mosquitoes and in agriculture to address pests damaging crops.

Advocates for pesticide use emphasize its pivotal role in enhancing agricultural productivity. The adoption of pesticides has contributed significantly to a fourfold increase in food grain production in India, from 50 million tons in 1948–1949 to 198 million tons by the end of 1996-1997. With the world's growing population, many consider this a good thing. Pesticides also play a crucial role in reducing losses from weeds, diseases, and insect pests, contributing to a substantial increase in harvestable produce. For instance, in rice cultivation, weed control practices have been shown to prevent produce reductions from weeds ranging from 28% to 48%. In dry land crops, weeds can reduce produce by 37–79%, and herbicides offer both economic and labor benefits by mitigating these losses (Aktar et al., 2009). Furthermore, pesticides, particularly insecticides, play a crucial role in controlling vector-borne diseases, such as malaria, by targeting and killing disease-carrying vectors. Moreover, pesticides contribute to ensuring food quality by protecting crops from pests and diseases, ultimately enhancing nutritional properties (Aktar et al., 2009).

But it is important to look at the big picture. Detractors note that using pesticides might pose health risks, which could lead to neurological diseases (Kamel & Hoppin, 2004), reproductive issues (Fucic et al., 2021), and other long-term health problems. Moreover, their usage might lead to disrupting endocrine systems, causing hormonal disruptions, diminished intelligence, and cancer (Californians for Pesticide Reformer, n.d.). On top of that, they are one of the main culprits of groundwater contamination, with 143 different pesticides being found in the groundwater of 43 US states. Herbicides and fungicides can move through water and impact the quality of nearby water sources. Similarly, pesticides can harm soil by reducing fertility, altering soil pH, and eliminating beneficial microorganisms essential for healthy plant growth (NPIC, 2023). The impact on fungi and bacteria shows that pesticides can disrupt soil ecosystems, leading to long-term degradation (NPIC, 2023).

You own a conventional farm that uses minimal synthetic chemicals but does not qualify as organic. A pesticide company has approached you with an offer to supply pesticides at a discounted rate.

- Would you agree to use pesticides on your farm? Why or why not?
- Explain and justify your decision.
- Your neighbor, who runs an organic farm, disagrees with you. What does she say to convince you of her position?
- How would you defend your decision?
- How do scientists study the effects of pesticides? (Consider methods such as long-term studies, controlled experiments, and field research.)
- Why might scientists reach different conclusions even when examining the same data about pesticides?
- Do you think the knowledge about pesticides might change in the future? Why or why not?

31. Acid Rain

Since the 1980s, acid rain has been a significant environmental concern, particularly with the acidification of various surface waters worldwide. Although the issue has diminished in many regions due to successful sulfur emission control programs in Europe and North America, its effects persist in certain areas. The primary cause of acid rain is the atmospheric deposition of sulfur and nitrogen oxides, which react with water to form diluted sulfuric and nitric acids. These acids can lower the pH of lakes and streams, leading to long-term environmental damage.

The expectation was that affected aquatic ecosystems would recover following sulfur emission reductions, but progress has been slower than anticipated. Despite substantial reductions in sulfur emissions, the chemical recovery of acidified lakes has been limited outside specific regions like Sudbury, Canada, where aggressive sulfur control measures were implemented. Models suggest that even with improvements in emission controls, lakes in some countries may continue to face acid damage.

Advocates for current sulfur emission reduction strategies argue that extensive biological recovery is possible when chemical conditions in lakes improve, though not all species can recover fully. While communities in recovered lakes may not return to pre-disturbance conditions, ongoing measures, such as reducing sulfur emissions from coal-burning power plants, have significantly decreased atmospheric sulfur levels. These programs have led to drastic falls in emissions, but challenges remain, such as persistent chemicals in lakes and agricultural damage linked to acid rain.

Critics of current approaches to sulfur reduction point out unintended consequences. For example, emissions from coal combustion not only release sulfur dioxide but also particulate matter, which contributes to air pollution and health risks, including cancer and asthma. These particles can travel long distances and deposit into freshwater systems, further impacting water quality. Alternative solutions, such as liming, have been proposed to address the acidification of lakes. Liming involves adding calcium carbonate or other alkalines to neutralize water pH, offering a complementary approach to sulfur emission controls.

- Do you support the current efforts to mitigate acid rain and reduce sulfur emissions?

Yes No

- Provide an explanation for your chosen position.
- How would you balance the legal and environmental implications of lawsuits related to the aftermath of acid rain, especially concerning damage to agricultural land?
- How can you explain that scientists reached different conclusions even though scientists were all looking at the same data about the effects of acid rain?
- How do scientists know about the effects of acid rain?
- How certain are scientists about the knowledge related to acid rain?
- Do you think the knowledge about acid rain might change in the future? Explain why or why not.

32. **Cool Seal**

To address the challenges of climate change, the city of Los Angeles initiated a groundbreaking experiment in 2015, employing a novel technique known as Cool Seal. Developed by the company GuardTop, Cool Seal involves spraying streets with a reflective coating that has a greyish-white color to combat the urban heat island effect, reflecting more sunrays and consequently cooling down the city's streets. The application process utilizes trucks and squeegees to cover the surfaces with this reflective material. Los Angeles, like many other cities, is predominantly covered with thousands of kilometers of asphalt. Research indicates that traditional dark-colored asphalt absorbs between 80% and 95% of the sun's rays, significantly heating up streets and the surrounding areas. As temperatures in Southern California soar above 37 °C, the asphalt roads can reach temperatures of approximately 65 °C. This intensifies the "Urban Heat Island Effect," contributing to a 5 °C increase in the average temperature of Los Angeles compared to nearby regions, as reported by the Environmental Protection Agency. Cool Seal has emerged as a solution, but not everyone views it that way.

Advocates of Cool Seal argue that it helps bounce off solar rays from asphalt, reducing heat absorption. A study conducted by the Bureau of Street Services found that streets covered with lighter colors using Cool Seal were on average 9 °C–12 °C cooler than non-covered areas. This not only decreases the need for air conditioning but also has potential environmental benefits.

However, a critical examination is essential. While the decrease in surface temperature has been noted by city officials, the impact on air temperature and human health needs consideration. In a study conducted by UCLA urban planners, unshaded streets in Pacoima and Sun Valley were examined for air temperature, wind speed, humidity, and radiation. The results revealed alarming consequences, with reflected roadways increasing temperatures by 31 degrees on a typical summer day. The mean radiant temperature, a measure of heat radiated by surrounding surfaces, was over 14 degrees higher on the painted pavements, with this elevated temperature persisting throughout the day and increasing at night. According to bio climatologist Larry Kalkstein, the increased heat may have adverse effects on human health, leading to more sweating, increased water consumption, and potential respiratory distress. The vulnerable groups, including the elderly, homeless, and individuals with existing health conditions, could face heightened risks.

- Considering the potential benefits and drawbacks of Cool Seal in combating the urban heat island effect, do you believe the city of Los Angeles should continue its use of this reflective coating on streets?

Yes No

- Explain and justify your position.
- How do scientists know about the effects of cool seal?
- Do you think your decision might change in the future? Why or why not? Explain.
- Why did scientists reach different conclusions even though they were looking at the same data?

33. **Cloud Seeding**

Cloud seeding is a way of making man-made clouds that can provide shade. This is achieved by dispersing substances in the air that act as ice nuclei or condensation. The debate over cloud seeding centers on its potential benefits for water resources and agriculture versus the environmental and economic drawbacks associated with the chemicals used and the high costs of implementation.

Some scientists argue that cloud seeding can increase rainfall by 10–15%, offering a potential solution to water scarcity issues, particularly in drought-prone areas like the UAE. Countries supporting cloud seeding emphasize its economic benefits, suggesting that it can enhance agricultural development by increasing crop yield and minimizing damage to crops caused by atmospheric changes like heat.

However, opposing views suggest that the use of harmful chemicals, such as silver iodide, in cloud seeding poses significant environmental and health risks. Detrimental effects range from potential floods to undesirable weather patterns.

Additionally, the process is considered expensive and demands substantial investments, which could have long-term economic implications for the country.

- Would you vote for applying cloud seeding in your country?

YES NO

- Explain and justify your decision.
- Another scientist, Professor Ponso, disagrees with your position. How could he explain his position to illustrate the reasons supporting it and convince you?
- What would you explain to Professor Ponso to explain that your position is right?
- How can you explain that scientists reached different conclusions even though scientists were all looking at the same data about the effects of cloud seeding?
- How do scientists know about the effects of cloud seeding?
- How certain are scientists about the knowledge related to cloud seeding?
- Do you think the knowledge about cloud seeding might change in the future? Explain why or why not.

34. **Quarrying**

Quarrying involves the extraction of geological materials from the Earth. This practice has been undertaken for millennia, playing a pivotal role in human development, particularly in construction and architecture. In Lebanon, many quarries focus on extracting limestone from hills and mountains to produce stone slabs and blocks. This activity has evolved into a socio-scientific and environmental issue that has gained significant attention and has been politicized.

Supporters of quarrying contend that it can be carried out sustainably and in a regulated manner. They point to the National Master Plan for Quarries established by the Council of Ministers and the identification of twenty areas by the Directorate General of Urban Planning (DGUP) that can be quarried without causing direct environmental harm. Advocates also emphasize the economic benefits, asserting that local income is generated through the extraction, production, and sale of stone. This, they argue, facilitates easier construction, reduces the need for stone imports, and subsequently lowers costs. Proponents further claim that sustainable quarrying practices are feasible, especially in areas approved by the DGUP.

On the opposing side, critics of quarrying express concerns about the loss of biodiversity and habitats. One major worry is the impact on Lebanon's endemic Mediterranean species of fauna and flora. A 2008 study revealed that, "in 2005, 21.5 percent were in forests and fertile lands, 32.4 percent were concentrated in scrubland and grassland, and 3.2 percent were located in urban areas" (Public Works Studio, 2019). The loss of vegetation results in habitat destruction, jeopardizing the entire ecosystem and increasing the risk of desertification, which poses threats to both animals and humans in the long run. Additionally, the dust produced by quarries adversely affects people's health as they inhale particles. Critics also point out that the transportation of limestone from quarries to production sites contributes to CO_2 and carbon monoxide emissions from trucks, exacerbating environmental issues, albeit to a lesser degree than other quarrying impacts. Finally, it is asserted that many quarries in the country are not approved by the National Physical Master Plan of the Lebanese Territory (NPMPLT) and are situated outside authorized areas.

Suppose you are a Lebanese citizen. Lebanon plans on creating more quarries to supply stone for local construction companies and requires Lebanese citizens to vote for or against this issue.

- Would you vote for adding more quarries in Lebanon?
- Explain the reasons behind your answer.
- The town mayor and the NPMPLT were opposed to the decision. What are their possible arguments?
- In what ways would you convince them otherwise. Explain using reasons to back up your answer.
- How can you explain the polemic behind such a topic?
- Do you think that the development of quarries is beneficial to the socio-economic realm? Or is it deemed more harmful in the long run?
- Do you think further research should be done on that subject?
- Do you think you might change your position on that matter in the future?
- Do you think there is a way of quarrying in a sustainable manner that does not harm biodiversity?

35. **Global Warming**[2]

Global warming is a major environmental issue facing the international community.

According to the majority of scientists, human activities, especially the burning of fossil fuels (oil, gas, and coal), have significantly increased the levels of carbon dioxide and other greenhouse gases in the atmosphere. These gases trap solar energy, which raises the Earth's temperature and may lead to an environmental disaster.

According to a minority group of scientists, the Earth is going through natural warming cycles. They argue that the influence of humans is insignificant since temperature increases are part of the Earth's natural climate history, as evidenced by alternating ice ages and warmer periods. This group also worries that proposed "solutions" to global warming could have significant negative effects on the global economy.

In 1997, a climate conference in Kyoto (Japan) brought together representatives from more than 160 nations to discuss binding limitations on greenhouse gases. The outcome of the meeting was the Kyoto Protocol, which required developed countries to commit to reducing their global emissions by 5.2% by 2012 (Grubb et al., 2002). Developing countries were exempted from legally binding commitments but were encouraged to reduce their emissions with financial and technical support. So far, about 140 countries have agreed to abide by the Kyoto Protocol. The United

[2]This scenario was first published in Khishfe, R. (2022). Nature of science and argumentation instruction in socioscientific and scientific contexts. *International Journal of Science Education, 44*(4), 647–673. doi:https://doi.org/10.1080/09500693.2022.2050488

States did not ratify the protocol, citing concerns about its economic impact and the lack of binding commitments for rapidly developing countries like China and India. However, the US has joined other environmental treaties.

It needs to be noted that the restrictions imposed by the protocol contributed to significant reductions in CO_2 emissions in many countries. However, some sectors experienced economic challenges, including slower GDP growth. At the same time, the protocol encouraged investment in greener technologies, and some predict that economies will gradually adapt to these shifts.

Research indicates that developing countries should also limit emissions as their emissions are predicted to exceed those of developed countries in 10 years (Kim et al., 2020).

- • Do you think that the US should have joined the Kyoto Protocol?

YES NO

- Do you think that the US should agree to be legally responsible for limiting greenhouse gas emissions and reducing them to prevent dangerous climate change? Explain and justify your decision.
- Another scientist, Professor Ponso, disagrees with your decision. How could he explain his position to illustrate the reasons supporting it and convince you?
- What would you reply to Professor Ponso to explain that your decision is right?
- How can you explain that scientists reached different conclusions even though they were all looking at the same data about the possible outcomes of global warming?
- How do scientists know about the effects of global warming? Provide examples of scientific methods or research that have helped establish this knowledge.
- How certain are scientists about the knowledge related to global warming? Highlight areas of high confidence and where uncertainties remain.
- Do you think the knowledge about global warming might change in the future? Explain why or why not.
- As a scientist, do you think you might change your decision in the future? Explain why or why not.

References

Abbas, I., Chaaban, J., Al-Rabaa, A., & Shaar, A. (2019). Solid waste management in Lebanon: Challenges and recommendations. *Journal of Environment and Waste Management, 4*(2), 53–63. Retrieved from https://www.researchgate.net/publication/331378505_Solid_Waste_Management_in_Lebanon_Challenges_and_Recommendations_JEWM_Solid_Waste_Management_in_Lebanon_Challenges_and_Recommendations

Aktar, M. W., Sengupta, D., & Chowdhury, A. (2009, March). Impact of pesticides use in agriculture: Their benefits and hazards. *Interdisciplinary Toxicology, 2*, 1–12. https://doi.org/10.2478/v10102-009-0001-7

Andersen, L. E. (2002). *The dynamics of deforestation and economic growth in the Brazilian Amazon*. Cambridge University Press.

Aoun, A., Darwiche, F., Hayek, S. A., & Doumit, J. (2018). The fluoride debate: The pros and cons of fluoridation. *Preventive Nutrition and Food Science, 23*(3), 171–180. https://doi.org/10.3746/pnf.2018.23.2.171

Bailis, J. A., Ezzati, M., & Kammen, D. M. (2005). Mortality associated with indoor air pollution from biomass fuel use in rural China. *Environmental Health Perspectives, 113*(7), 997–1002. https://doi.org/10.1289/ehp.7967

Bassin, E. B., Wypij, D., Davis, R. B., & Mittleman, M. A. (2006). Age-specific fluoride exposure in drinking water and osteosarcoma (United States). *Cancer Causes & Control, 17*(4), 421–428. https://doi.org/10.1007/s10552-005-0500-6

Bodetti, A. (2019, May 2). *Can recycling save Lebanon?* Lobe Log. Retrieved from https://lobelog.com/can-recycling-save-lebanon/

Bou-Orm, I. R., Sakr, R. E., & Adib, S. M. (2018). Cervical cancer screening among Lebanese women. *Revue d'Épidémiologie et de Santé Publique [The Journal of Epidemiology and Public Health], 66*(1), 1–6. https://doi.org/10.1016/j.respe.2017.10.004

Californians for Pesticide Reform. (n.d.). *Pesticides and human health*. Retrieved April 12, 2024, from https://www.pesticidereform.org/pesticides-human-health/#:~:text=Pesticides%20have%20been%20implicated%20in,spontaneous%20abortion%2C%20sterility%20and%20infertility

Courtat, C., Khalife, A., & Mendret, M. (2023, June 12). "The coast is ours! This is a theft!": Beach privatization provokes protests. *L'Orient Today*. Retrieved from https://today.lorientlejour.com/article/1340270/the-coast-is-ours-this-is-a-theft-beach-privatization-provokes-protests.html

Dagher, J. (2020). *Urbanization and urbanicity association with mental health and psychological well-being: A case study in Lebanon* (Master's thesis, Politecnico di Torino). Webthesis Libraries. Retrieved from https://webthesis.biblio.polito.it/16607/

Fawaz, M. (2017, May 1). *Urban policy: A missing government framework*. The Lebanese Center for Policy Studies. Retrieved from https://www.lcps-lebanon.org/articles/details/1917/urban-policy-a-missing-government-framework

Ferronato, N., & Torretta, V. (2019). Waste mismanagement in developing countries: A review of global issues. *International Journal of Environmental Research and Public Health, 16*(6), 1060. https://doi.org/10.3390/ijerph16061060

Fucic, A., Duca, R. C., Galea, K. S., Maric, T., Garcia, K., Bloom, M. S., Andersen, H. R., & Vena, J. E. (2021). Reproductive health risks associated with occupational and environmental exposure to pesticides. *International Journal of Environmental Research and Public Health, 18*(12), 6576. https://doi.org/10.3390/ijerph18126576

Germanos, P. S., & Azzi, S. (2024, January 8). *The devastating impact of Lebanon's environmental failures*. The Washington Institute. Retrieved from https://www.washingtoninstitute.org/policy-analysis/devastating-impact-lebanons-environmental-failures#:~:text=Air%20Pollution%3A%20Lebanon's%20air%20quality,issues%20in%20humans%20and%20animals

Grubb, M. J., Hope, C., & Fouquet, R. (2002). Climatic implications of the Kyoto Protocol: The contribution of international spillover. *Climatic Change, 54*(1–2), 11–28. https://doi.org/10.1023/A:1015775417555

Hourany, D. (2023, June 27). *Growing lack of accessibility to beaches in Lebanon*. Now Lebanon. Retrieved from https://nowlebanon.com/growing-lack-of-accessibility-to-beaches-in-lebanon/

Hu, S. (2022, October 13). *Ocean acidification: What you need to know*. NRDC. Retrieved from https://www.nrdc.org/stories/ocean-acidification-what-you-need-know.

Institute of Medicine. (2011). *Dietary reference intakes for Calcium and Vitamin D*. National Academies Press. https://doi.org/10.1210/jc.2010-2704

Kamel, F., & Hoppin, J. A. (2004). Association of pesticide exposure with neurologic dysfunction and disease. *Environmental Health Perspectives, 112*(9), 950–958. https://doi.org/10.1289/ehp.7135

Keuchkerian, K. (2022, July 22). *Here are the 13 most polluted and unsafe beaches in Lebanon*. The 961. Retrieved from https://www.the961.com/13-most-polluted-unsafe-beaches-lebanon/

Khalil, S. (2022, September 15). *The state of recycling in Lebanon: Q&A with waste management expert Samar Khalil*. Berytech. Retrieved from https://berytech.org/the-state-of-recycling-in-lebanon-qa-with-waste-management-expert-samar-khalil/

Khamis, R. (2018, October 10). *Urbanization in Lebanon: A problematic local issue, reflecting global trends*. Annahar. Retrieved from https://www.annahar.com/english/article/877425-urbanization-in-lebanon-a-problematic-local-issue-reflecting-global-trends

Khan, I., Zakari, A., Zhang, J., Dagar, V., & Singh, S. (2022). A study of trilemma energy balance, clean energy transitions, and economic expansion in the midst of environmental sustainability: New insights from three trilemma leadership. *Energy, 248*, 123619. https://doi.org/10.1016/j.energy.2022.123619

Kim, Y., Tanaka, K., & Matsuoka, S. (2020). Environmental and economic effectiveness of the Kyoto Protocol. *PLoS One, 15*(7), e0236299. https://doi.org/10.1371/journal.pone.0236299

Laurance, W. F., Williamson, G. B., Laurance-Silva, P. R., et al. (2022). An empirical assessment of the impact of Amazonian indigenous and protected areas on forest loss and fire. *Proceedings of the National Academy of Sciences, 119*(36), e2209513119. Retrieved from https://www.nytimes.com/2022/03/16/magazine/amazon-rainforest-ituna-itata.html

Ma, H., Nittrouer, J. A., Fu, X., Parker, G., Zhang, Y., Wang, Y., Wang, Y., Lamb, M. P., Cisneros, J., Best, J., Parsons, D. R., & Wu, B. (2022). Application of downstream flood stage due to damming of fine-grained rivers. *Nature Communications, 13*, 3054. https://doi.org/10.1038/s41467-022-30730-9

Massoud, M., Merhebi, F., Geha, C., Prattis, S., Rizk, N., Saliba, N. A., Talhouk, S., Mroweh, A., & Yakzan, S. (2016). *Guide to municipal solid waste management*. American University of Beirut. Retrieved from https://www.aub.edu.lb/natureconservation/Documents/guide_to_municipal_solid_waste_management.pdf

McGrath, M. (2019, July 3). *US top of the garbage pile in global waste crisis*. BBC News. Retrieved from https://www.bbc.com/news/science-environment-48838699

Naharnet Newsdesk. (2014, December 11). *A garden blooms on Sidon's former 'trash mountain'*. Naharnet. Retrieved from https://www.naharnet.com/stories/en/158710

National Geographic Education. (n.d.). *Earth's fresh water*. Retrieved April 12, 2024, from https://education.nationalgeographic.org/resource/earths-fresh-water/

National Pesticide Information Center. (2023). *Pesticides and soil health*. Retrieved from https://npic.orst.edu/envir/soil.html

Public Works Studio. (2019, April 17). *Reading the quarries' map in Lebanon*. Jadaliyya. Retrieved April 11, 2022, from https://www.jadaliyya.com/Details/38569

Ritchie, H. (2018). *How urban is the world?* Our World in Data. Retrieved from https://ourworldindata.org/how-urban-is-the-world#:~:text=although%20it%20can%20seem%20like,defined%20as%20built%2Dup%20area%3B&text=rates%20of%20urbanization%20have%20been,lived%20in%20urban%20areas)%3B

Ritchie, H., Samborska, V., & Roser, M. (2024). *Urbanization*. Our World in Data. Retrieved from https://ourworldindata.org/urbanization

Sbeih, R. (2023). *The case study of transitioning centralized solid waste management in Lebanon: The devolution and its limitations* [Master's thesis, University of Groningen]. Student Theses. Retrieved from https://frw.studenttheses.ub.rug.nl/4162/

Shaban, A. (2019). Striking challenges on water resources of Lebanon. In M. S. Javaid (Ed.), *Hydrology - The science of water*. IntechOpen. https://doi.org/10.5772/intechopen.81878

Sofian, L. (2023, August 28). *Trash talk: How a local initiative is helping to clean up Lebanon*. The New Arab. Retrieved from https://www.newarab.com/features/swim-meet-trashy-local-initiative-cleaning-lebanon

Till, C., & Green, R. (2021). Controversy: The evolving science of fluoride: When new evidence doesn't conform with existing beliefs. *Pediatric Research, 90*(5), 1093–1095.

UN-Habitat. (2011). *Lebanon urban profile: A desk review report*. Retrieved from https://unhabitat.org/sites/default/files/download-manager-files/Lebanon%20Urban%20Profile.pdf

United Nations Development Programme. (2020). *Energy and waste solutions*. Retrieved April 12, 2024, from https://www.undp.org/lebanon/projects/energy-and-waste-solutions-0

United Nations Food and Agriculture Organization. (2005). *FRA 2005 maps and figures*. Retrieved from https://www.fao.org/forestry/fra/41256/en/

United Nations Food and Agriculture Organization. (2023a). *World agricultural production: A mid-term analysis*. Retrieved from https://www.fao.org/3/cc3751en/cc3751en.pdf

United Nations Food and Agriculture Organization. (2023b). *World database of protected areas*. Land & Water. Retrieved from https://www.fao.org/land-water/land/land-governance/land-resources-planning-toolbox/category/details/en/c/1236440/

Warren, C., & Steenbergen, D. J. (2021). Fisheries decline, local livelihoods and conflicted governance: An Indonesian case. *Ocean & Coastal Management, 202*, 1–13. https://doi.org/10.1016/j.ocecoaman.2020.105498

World Wide Fund for Nature-Brazil. (2022, November 30). *Deforestation in the Amazon remains at high levels with a rate of 11,568 km^2 in 2022*. Retrieved from https://www.wwf.org.br/?84261/Deforestation-in-the-Amazon-remains-at-high-levels-with-a-rate-of-11568-km2-in-2022#:~:text=The%20annual%20rate%20of%20deforestation,for%20Space%20Research%20(INPE)

Youssef, S. (2019). *Women's rural-urban migration in Lebanon: Migration determinants and the quest for empowerment* [Master's thesis, Lebanese American University]. Lebanese American University Repository. Retrieved from https://laur.lau.edu.lb:8443/xmlui/bitstream/handle/10725/12298/Sally_N_Youssef_Thesis_Redacted.pdf?sequence=1&isAllowed=y

Chapter 6
Chemistry

Abstract This chapter investigates the environmental and health consequences of different chemical-related topics, such as electric vehicles, plastic and air pollution, lithium mining, hydraulic fracturing, fossil fuel burning, and the safety of chemicals in cosmetic goods. While electric vehicles are frequently viewed as a sustainable alternative to gasoline-powered vehicles, questions regarding battery manufacture, raw material extraction, and electrical sources call into question their environmental friendliness. Similarly, plastic waste poses a danger to ecosystems and human health, with continuous disputes about how to reduce plastic usage while maintaining economic advantages. Air pollution, caused by both human and natural activity, continues to be a major worldwide issue, necessitating policy considerations. Lithium mining and hydraulic fracturing are being scrutinized for their environmental consequences, balancing energy requirements with ecological impact. The burning of fossil fuels, which contributes to acid rain, sparks discussions over regulatory actions. Finally, the existence of potentially hazardous compounds in cosmetic products raises questions regarding consumer safety and governmental monitoring. The chapter gives opposing viewpoints on these problems, encouraging critical dialogue on sustainability, scientific consensus, and the changing nature of environmental knowledge.

Keywords Environmental pollution · Electric vehicles · Plastic waste · Fossil fuel · Chemical safety

This chapter delves into socioscientific issues in the field of chemistry, concentrating on the environmental consequences of critical themes. Beginning with an investigation of electric cars and their relationship to pollution, how these vehicles may contribute to environmental issues is presented. The focus then shifts to electric cars, with an emphasis on their role in reducing pollution. Moving on to the prevalent issue of plastic pollution, an exploration is presented on the environmental consequences and the complications that arise from our reliance on synthetic materials. Furthermore, the chapter dives into the larger topic of air pollution, revealing the chemical composition of contaminants and their possible influence on human health.

R. Khishfe, *The Complexities of Socioscientific Issues in Education*, Springer Texts in Education, https://doi.org/10.1007/978-3-031-88006-3_6

1. Electric Vehicles and Pollution
2. Electric Vehicles
3. Plastic Pollution
4. Air Pollution
5. Lithium Mining
6. Fracking and Hydraulic Fracturing
7. The burning of fossil fuels
8. Behind the Mask: The Debate About Chemicals in Beauty Products

1. Electric Vehicles and Pollution

Pollution is a global environmental issue that has been increasing significantly over the years. It results from the release of chemical substances and toxins into the environment, negatively impacting ecosystems and people's health. The persistent high levels of pollution have become a cause for concern, leading many countries worldwide to act, including the development of alternatives to traditional sources of pollution. One such invention is the electric vehicle (EV), which first emerged in response to the oil crisis and growing climate concerns in the late twentieth century.

One group of people supports electric vehicles. In fact, many countries have embraced electric vehicles due to their numerous advantages. EVs produce no direct CO_2 emissions during operation. However, their overall carbon footprint depends on the energy source used for charging. Electricity generated from renewable sources significantly reduces emissions, while reliance on fossil fuels for power generation can still contribute to greenhouse gas emissions. Additionally, operating EVs is more cost-effective overall. This is largely because EVs do not have combustion engines and require fewer oil changes, fewer mechanical repairs, and less frequent part replacements compared to gas-powered vehicles. Moreover, EVs are highly energy-efficient, with efficiency ratings of around 70%, in stark contrast to gasoline engines, which typically offer efficiencies of only about 15%.

However, it is essential to consider both the positive and negative impacts of using electric vehicles. Detractors state that even though EVs have advantages, their manufacturing process, especially the production of batteries, requires the extraction of raw materials such as lithium and nickel, contributing to water pollution and carbon emissions. The environmental impact of EVs is also influenced by the source of electricity used for charging. If electricity is generated from fossil fuel-based power plants like coal and natural gas, EVs can still contribute to greenhouse gas emissions. Lastly, battery disposal is an issue that needs more attention, as batteries have a limited lifespan, and a proper recycling and disposal system is yet to be fully established.

- Do you believe the environmental benefits of electric vehicles, such as reduced carbon emissions and increased energy efficiency, outweigh the negative impacts associated with their manufacturing, reliance on raw materials, and potential contributions to pollution based on the source of electricity used for charging?

Yes No

- Explain and justify your position.

- Another scientist, Professor Ponso disagrees with your position. What would he say to convince you of his stance?
- How would you defend your original viewpoint?
- Do you think your decision might change in the future? Why or why not? Explain.
- Why did scientists reach different conclusions even though they were looking at the same data?

2. Electric Vehicles

An electric vehicle (EV) is a car that operates on electricity, either entirely or partially. EVs are known for their minimal operating expenses due to having fewer moving parts and their eco-friendly nature as they consume minimal fossil fuels, such as petrol or diesel. There are various perspectives on the emergence of EVs in today's world, with many viewing it as a positive, as well as others viewing it as less advantageous.

The first group explains that one primary benefit is the ability to lower greenhouse gas emissions, a significant environmental advantage. Traditional gas-powered cars contribute substantially to greenhouse gas emissions, with transportation accounting for 29 percent in 2019, according to the U.S. Environmental Protection Agency (EPA). Electric cars, lacking an exhaust pipe, do not emit greenhouse gases, although the environmental impact depends on the energy source used. If generated from renewable sources like wind, water, or solar energy, electricity becomes a cleaner alternative compared to finite resources like oil, used in gas-powered vehicles.

Despite the initial appeal of electric vehicles, closer scrutiny reveals potential drawbacks. People against electric vehicles, particularly cite concerning carbon footprints and the extraction of metals like lithium and cobalt. The production of electric cars entails emissions, starting with the mining of basic materials, which emit greenhouse gases. The raw materials must undergo processing, releasing more emissions, and the manufacturing process further contributes to carbon emissions. Charging EVs poses another environmental challenge, especially when the electricity comes from coal-fired power stations. Thus, the overall environmental friendliness of electric vehicles depends on the energy source used for both manufacturing and charging.

Your city is planning on generally changing all gas vehicles with EV in a chance to better environmental causes. What is your take on it?

- Would you vote for integrating EVs in the city and stopping gas vehicle manufacturing?

Yes No

- Explain the reason behind your answer.
- A town consultant is against the idea of replacing all gas vehicles. What would their argument be?
- What would you say to them?
- Would changing all cars make it way better environmentally?
- Would you think more research should be done so we could decide?

- Do you think you might change your decision in the future?

3. Plastic Pollution

Due to its pervasiveness and harmful impacts on ecosystems and human health, plastic pollution has become a critical global environmental issue. The problem has spurred discussions and disputes about its causes, effects, and potential solutions.

The group against using plastic argues that plastic has a great impact on the environment. It takes hundreds of years to decompose, causing harm to the oceans and marine life, as well as the land and terrestrial animals. "Plastic is persistent in the environment, with rates of natural removal on the scale of decades to centuries" (MacLeod et al., 2021). Furthermore, they highlight that plastics can harm humans as well. Microplastics can penetrate the human body through ingestion. "Microplastics can be ingested by low trophic suspension […]. Therefore, they may accumulate within organisms, resulting in physical harm" (Wright et al., 2013).

On the other hand, another group recognizes the detrimental impacts of plastic pollution but raises concern about the material benefits. "Plastics can be extremely lightweight and strong. Consequently, less energy is used during the transportation of the packaged products. In addition, less energy and water are required during the manufacture of plastic packaging" (Bernau, 2019). They also argue that the plastic industry contributes to economic growth. "The plastics industry is one of the largest manufacturing industries in the U.S. It generates $451 billion per year in shipments. It employs 993,000 people directly and more than 1.5 million indirectly."

Your city plans on addressing plastic pollution through consumer awareness and behavior change. It requires residents to vote for or against this issue.

- Would you vote for the reduction of the production and consumption of single-use plastic?

Yes No

- Explain and justify your decision.
- Should the city adopt tighter laws and guidelines to restrict the manufacture, sale, and use of single-use plastics?
- Would you support providing financial incentives or awards to people or organizations who actively minimize their use of plastic and switch to more eco-friendly options?
- What types of campaigns or activities do you believe would be successful in raising awareness about plastic pollution and encouraging behavior change?
- How do scientists know about the effects of plastic pollution?
- Do you think your decision might change in the future? Why or why not? Explain.
- Why did scientists reach different conclusions even though they were looking at the same data?

4. Air Pollution

The presence of harmful substances in the atmosphere which cause the contamination of air is known as air pollution. Air pollution is a growing problem worldwide, with respiratory diseases on the rise. While everyone can agree that it is a problem that affects a lot of countries, there are two opposing groups when it comes to the causes of air pollution.

The first perspective indicates that some human actions are the cause of air pollution. This can be seen on a national level. For example, during the revolution in Lebanon, protestors used to burn wheels which lead to the emission of smoke in the air and polluting it. Most Lebanese adults smoke more than one packet of cigarette a day. Moreover, smoke exhausted by companies, industry factories, and vehicles are causes of air pollution as well. All these human activities harm the atmosphere and cause fatal diseases such as lung cancer and cardiovascular illnesses.

The counterargument's point of view states that air is polluted naturally. Forest fires that happened naturally in Lebanon cause the air to pollute due to the smoke. Additionally, some natural phenomena like earthquakes and volcanic eruptions affect air quality. Finally, plant-based organic compounds like volatile organic compounds (VOCs) harm the atmosphere. Studies show that "some people suffer from lung cancer due to VOCs."

Suppose you live in Lebanon, and that the government has just imposed a fine on smoking in order to mitigate environmental and health risks.

- Do you agree with the policies set to prevent people from smoking?

Yes No

- Do you think the government of Lebanon could have taken other measures to solve the issue of air pollution? Give some examples.
- Some scientists claim that air pollution is caused by nature and humans have nothing to do with it. What would their arguments be to convince you?
- How would you reply to these scientists? Justify your answer.
- Another scientist states that air pollution in Lebanon is because of the government's mismanagement, do you agree?

Yes No

- Explain and justify your position.
- If you were a smoker, and those policies were put in place, how would you react? would you pay the fines and keep smoking? Explain.
- How do scientists know about the effects of air pollution?
- Do you think your decision might change in the future? Why or why not? Explain.
- Why did scientists reach different conclusions even though they were looking at the same data about the causes of air pollution?

5. Lithium Mining

Lithium mining is the extraction and processing of lithium, a soft, light, silvery-white alkali metal that has high electrochemical reactivity. As such, it is ideal for usage in batteries, making it an eco-friendly way to generate and store energy. But its extraction and processing have been mired with controversy.

For starters, supporters of lithium mining argue that it is necessary for the sake of transitioning into clean energy. After all, we would depend on lithium batteries for most of our activities, including using our laptops, cellphones, and even cars. It could also be used to store wind and solar energy. All of this would help us reduce our reliance on fossil fuels, and thus decrease greenhouse gas emissions and global warming. Thus, this group sees that the end justifies the means.

On the other hand, there is a group of people against lithium mining. Those argue that sometimes, the lithium might need to be extracted from igneous rocks with pegmatitic minerals. Some of the effects of hard rock mining include soil erosion, deforestation, and habitat destruction – all of which affect water supplies. In addition, this process could lead to changes in soil profiles, increased dust, and noise levels, directly affecting human health. Even the alternate process of extracting lithium via underground mineral water is less-than-desirable due to the resulting water pollution. Moreover, transportation, as well as electricity generation for the purpose of separating lithium from other elements, produce greenhouse gases.

- Do you agree with lithium mining?

Yes No

- Explain and justify your decision.
- Professor Nyong'o disagrees with you. What does he say to convince you of his position?
- How would you reply in defense of your stance?
- How do scientists know about the effects of lithium mining?

- Do you think your decision might change in the future? Why or why not? Explain.
- Do you think the knowledge about lithium mining might change in the future? Explain why or why not.

6. Fracking

Fracking, also known as hydraulic fracturing, is a technique used to extract natural gas and oil from deep underground rock formations by injecting a high-pressure mixture of water, sand, and chemicals into wells to fracture the rock and release the trapped hydrocarbons. This process allows access to previously inaccessible reserves and has led to a significant increase in domestic energy production. However, fracking is highly controversial due to concerns about its environmental impacts, including water contamination, air pollution, and seismic activity.

Advocates for hydraulic fracturing highlight its potential to unlock vast reserves of natural gas and oil, thereby reducing dependence on foreign energy sources and stimulating economic growth. They argue that fracking has led to a resurgence in domestic energy production, more jobs, revitalizing local economies, and generating revenue for governments through taxes and royalties. Additionally, they contend that natural gas, often touted as a cleaner-burning fossil fuel than coal, can serve as a bridge to a more sustainable energy future by reducing greenhouse gas emissions and facilitating the transition to renewable energy sources (Lenox & Kaplan, 2016).

Alternatively, critics of hydraulic fracturing raise concerns about its environmental and public health impacts, particularly regarding water and air quality. They point to incidents of water contamination from leaks, spills, and inadequate wastewater disposal practices associated with fracking operations. Moreover, they highlight the potential for air pollution from methane emissions, volatile organic compounds (VOCs), and hazardous air pollutants released during fracking (Weger et al., 2019). Additionally, opponents raise concerns about the risk of induced seismicity, or earthquakes triggered by the injection of fracking wastewater into underground disposal wells (Brudzinski & Kozłowska, 2019). They argue that the cumulative impacts of fracking pose significant risks to human health, ecosystems, and community well-being, warranting stricter regulation and oversight of the industry.

A gas company wants to use fracking on the borders of your city to extract energy. This could lead to more jobs, increased production, and thus, lower gas prices. Your city asks citizens to vote for this issue.

- Do you vote for or against the company's plans?
- Explain the reasoning behind your answer.
- Professor Ponso disagrees with your decision. What would he say to convince you of his position?
- How would you defend your original stance?
- How do scientists know about the effects of hydraulic fracturing?
- Do you think your decision might change in the future? Why or why not? Explain.

- Do you think the knowledge about hydraulic fracturing might change in the future? Explain why or why not.

7. The Burning of Fossil Fuels[1]

The burning of fossil fuels (especially in car engines and power stations) leads to the formation of pollutant gases that are acidic as sulfur dioxide and nitrogen dioxide. After these gases are released into the air, they react with rainwater causing rainwater to become acidic (acid rain). When acid rain falls into streams and lakes, it kills water life including fish. Additionally, acid rain may cause a negative effect on architectural buildings and works of art; it literally dissolves the structures that are constructed of marble (calcium carbonate) (Kaufman et al., 1991). There is a controversy on how to address the problem of acid rain.

The first group of scientists argues that the problem of acid rain is minor and that acidified lakes and streams can be treated by adding alkaline compounds (bases) to help restore the water to neutral conditions.

The second group argues that this solution is only temporary and if acid rain continues to fall, then bases must be reapplied. Instead, they strongly argue for a control on the release of pollutant gases in the air, which will reduce acid rain. This can be done by burning fewer fossil fuels and conserving more energy. Yet, that would lead to increases in the average cost of a new car, as well as special taxes to encourage energy conservation.

A third group argues for removing the pollutant gases before they get into the atmosphere. The extraction of these gases can also be very expensive, and again the people will have to pay for these costs.

There are three possible ways to address the problem of acid rain:

A. Not to do anything about acid rain because the problem is not very serious or can be solved easily.
B. Control the release of pollutant gases by burning less fossil fuel. Burning less fossil fuel means you would be prepared to go without a car of your own, or you would be prepared to have increases in the cost of a new car, and/or special taxes to encourage energy conservation.
C. Control the release of pollutant gases by removing them before they get into the atmosphere. Removing these gases can also be very expensive, and we will indirectly pay these costs.

- Which suggestion would you recommend?

A B C

- Explain and justify your decision
- Another scientist, Professor Ponso, disagrees with your decision. How could he explain his position to illustrate the reasons supporting it and convince you?

[1]This scenario was first published in Khishfe, R., Alshaya, F. S., BouJaoude, S., Mansour, N., & Alrudiyan, K. I. (2017). Students' understandings of nature of science and their arguments in the context of four socioscientific issues. *International Journal of Science Education, 39*(3), 299–334. https://doi.org/10.1080/09500693.2017.1280741.

- What would you reply to Professor Ponso to explain that your decision is right?
- How can you explain that scientists reached different conclusions even though they were all looking at the same data about the effects of acid rain?
- Do you think the knowledge about the effects of acid rain might change in the future? Please explain.
- Do you think you might change your decision in the future? Please explain.
- Is there anything else you would want to know about this issue that might help you decide or even change your decision?

8. Behind the Mask: The Debate About Chemicals in Beauty Products
Throughout history, beauty practices have used a variety of substances, from lead in Roman times for skin whitening (Olson, 2009) to modern formaldehyde lash glues (BL Lashes, n.d.). These substances highlight a long-standing issue: toxins and chemicals in beauty products. This has sparked public debate about the cost of beauty, gender norms, consumer safety, and environmental impacts. Many makeup products contain unsafe or unregulated substances, creating differing perspectives on their presence and the responsibility of beauty companies.

Supporters of the current state of makeup and beauty products argue that consumer safety is prioritized. These products go through rigorous testing by regulatory bodies such as the European Commission to ensure they are safe for consumers (European Commission, 2009). Major beauty brands, such as L'Oréal Groupe and Unilever, emphasize their commitments to sustainability and safety through consumer education and transparent business practices (L'Oréal Groupe, n.d.; Unilever, n.d.). Furthermore, research suggests that most of the chemicals used in beauty products are preservatives, which are essential to maintaining a product's freshness, texture, and longevity. According to (Lopez Muñoz et al., 2024), these preservatives support product health, consistency, and safety.

However, detractors raise serious concerns about the health and environmental consequences of these chemicals. Some ingredients in beauty products have been associated with skin irritation, hormone disruption, allergies, and other health problems (Sheikh, 2023). Exposure over time allows these chemicals to seep into the skin's pores, with effects that may not manifest for many years. Additionally, the production and disposal of these chemicals are environmentally harmful. They often end up in water systems, adversely affecting wildlife and ecosystems (Gutierrez, 2023).

Another growing concern is consumer trust. Many consumers report skepticism about companies' claims of safety and sustainability. According to Baker (2022), 79% of the public distrusts these promises, particularly because the Food and Drug Administration (FDA) does not actively regulate beauty product ingredients (Selby, 2023). This lack of oversight contributes to increased consumer unease, especially in light of chemical exposures and environmental degradation.

Your country is considering implementing stricter regulations on chemicals in beauty products to ensure consumer safety and environmental protection. This would involve increased oversight, mandatory labeling, and limits on certain chemical use. As a consumer, you recently read the ingredient list on one of your most frequently used beauty products and found several chemicals you've never heard of. This has left you questioning whether these substances are safe, whether companies can be trusted to disclose the risks, and whether environmental concerns should factor into your choice to continue using these products.

Now you are being asked to vote on whether your country should adopt stricter measures to regulate chemicals in beauty products.

- What do you vote for?
- Explain and justify your decision.
- Another expert, Professor John, disagrees with your position. What does he say to illustrate the reasons behind his stance and convince you?
- What would you reply to Professor John to explain that your position is right?
- How do scientists know about the effects of chemicals in beauty products?
- Do you think the knowledge about chemicals in beauty products might change in the future? Explain why or why not.

References

Baker, J. (2022, October 17). *79% of beauty shoppers have doubts about sustainability claims – How can brands rebuild trust?* Forbes. Retrieved from https://www.forbes.com/sites/jessibaker/2022/10/17/79-of-beauty-shoppers-have-doubts-about-sustainability-claims%2D%2Dhow-can-brands-rebuild-trust/

Bernau, P. (2019, August 12). *Plastic: Not as bad as often thought (KeyKeg, Trans.).* Retrieved from https://www.keykeg.com/news/plastic-not-bad-often-thought/#:~:text=Why%20is%20plastic%20so%20often,the%20manufacture%20of%20plastic%20packaging (Original work published 2019, June 22).

BL Lashes. (n.d.). *Eyelash extension glue ingredients: What's in my lash glue?* Retrieved July 17, 2024 from https://www.bllashes.com/blogs/blog/eyelash-extension-glue-ingredients

Brudzinski, M. R., & Kozłowska, M. (2019). Seismicity induced by hydraulic fracturing and wastewater disposal in the Appalachian Basin, USA: A review. *Acta Geophysica, 67*, 351–364. https://doi.org/10.1007/s11600-019-00249-7

European Commission. (2009). Regulation (EC) No 1223/2009 of the European Parliament and of the Council on cosmetic products. European Commission. https://eur-lex.europa.eu

Gutierrez, L. (2023, March 30). *Beauty at what cost? The environmental impact of cosmetic ingredients and materials*. Sustainable Earth. Retrieved from https://sustainable-earth.org/cosmetic-ingredients/#:~:text=These%20chemicals%20can%20enter%20the,be%20harmful%20to%20human%20health

Kaufman, Y. J., Fraser, R. S., & Mahoney, R. L. (1991). Fossil fuel and biomass burning effect on climate—Heating or cooling? *Journal of Climate, 4*(6), 578–588. https://doi.org/10.1175/1520-0442(1991)004<0578:FFABBE>2.0.CO;2

L'Oréal Groupe. (n.d.). *Green chemistry: The future of beauty*. Retrieved July 17, 2024, from https://www.loreal.com/en/articles/research-innovation/green-chemistry-the-future-of-beauty/

Lenox, C., & Kaplan, P. O. (2016). Role of natural gas in meeting an electric sector emissions reduction strategy and effects on greenhouse gas emissions. *Energy Economics, 60*, 460–468. https://doi.org/10.1016/j.eneco.2016.06.009

Lopez Muñoz, L. G., Barrera Rojas, L. M., Ospina Ladino, M. C., Beltran Jimenez, S. S., & Aguilar Ortiz, J. D. (2024). Determinación de la vida útil de un queso fresco con adición de aceite esencial de romero (*Salvia rosmarinus*) a partir del cumplimiento de la normativa colombiana vigente. *Investigación e Innovación en Ingenierías, 12*(2). https://doi.org/10.17081/invinno.12.2.7052

MacLeod, M., Arp, H. P. H., Tekman, M. B., & Jahnke, A. (2021). The global threat from plastic pollution. *Science, 373*(6550), 61–65. https://doi.org/10.1126/science.abg5433

Olson, K. (2009). Cosmetics in Roman antiquity: Substance, remedy, poison. *Classical World, 102*(3), 291–300. https://doi.org/10.1353/clw.0.0098

Selby, K. (2023, February 10). *45% of people worry about toxic makeup – Should you?* Asbestos. Retrieved from https://www.asbestos.com/featured-stories/makeup-toxicity-survey/

Sheikh, S. (2023). *Beauty care products - A chemistry deleterious to human chemistry.* ResearchGate. Retrieved from https://www.researchgate.net/publication/321322713_Beauty_care_products_-_A_chemistry_deleterious_to_human_chemistry.

Unilever (n.d.). Unilever's approach to sustainability. https://www.unilever.com/sustainable-living/

Weger, L. B., Lupascu, A., Cremonese, L., & Butler, T. (2019). Modeling the impact of a potential shale gas industry in Germany and the United Kingdom on ozone with WRF-Chem. *Elementa: Science of the Anthropocene, 7*, 49. https://doi.org/10.1525/elementa.387

Wright, S. L., Thompson, R. F., & Galloway, T. S. (2013). The physical impacts of microplastics on marine organisms: A review. *Environmental Pollution, 178*, 483–492. https://doi.org/10.1016/j.envpol.2013.02.031

Chapter 7
Interdisciplinary Science: Ethical and Societal Challenges

Abstract This chapter investigates the interaction of science, ethics, and society, addressing significant issues in several disciplines. Advances in artificial intelligence, remote work, and surveillance technologies are altering professional contexts, calling into question privacy, efficiency, and job security. Ethical quandaries in scientific research are highlighted, ranging from the argument over genetic alteration in people and crops to discussions over animal experimentation and cloning. The subject expands to sustainability and public health, examining the environmental implications of consumer behavior, the regulation of artificial food additives, and the ethical considerations associated with new health measures such as fat taxes. By deconstructing these interwoven concerns, the chapter emphasizes the rising necessity for responsible decision-making in science and technology, ensuring that innovation is consistent with ethical and social standards.

Keywords Science and ethics · Workplace technologies · Genetic modification · Artificial intelligence · Animal testing

This chapter investigates socioscientific scenarios, analyzing contemporary and ethical challenges in a variety of disciplines. In workplace dynamics, education, and society, the pros and cons of remote employment, the teaching style in Lebanon's public schools, and the issue of artificial intelligence "entering the job market" are explored. In ethical considerations in scientific research, the issues of animal experimentation, the use of animals for scientific or commercial research, the ethical implications of genetically engineered newborns, and the legality of euthanasia are presented. The investigation expands to genetic testing, including its uses, challenges, and social viewpoints, as well as ethical problems about changing legumes, adding chemical chemicals to crops, captive breeding in zoos, and cloning. Environmental and societal attitudes on science delves into the environmental challenge of rapid fashion, investigates societal attitudes on cannabis usage, and discusses the consequences of a "fat tax" on unhealthy foods in health and nutrition policies.

R. Khishfe, *The Complexities of Socioscientific Issues in Education*, Springer Texts in Education, https://doi.org/10.1007/978-3-031-88006-3_7

A. **Workplace Dynamics, Education, and Society:**

1. Balancing the Benefits and Drawbacks of Remote Work
2. Watched on the Job
3. Artificial Intelligence versus Human Intelligence
4. TikTok, Your Time Is Up
5. Aim Here for You: AI in Mental Health Applications
6. Gotcha: Facial Recognition and Privacy

B. **Ethical Considerations in Scientific Research:**

7. The Ethical Debate Over Animal Testing in Scientific Research
8. Human Subject Research
9. Genetically Modified Babies
10. Genetic Testing (Applications, Controversies, and Social Perspectives)
11. Modifying Legumes: Adding Vitamin D to tomatoes
12. Adding Chemical Substances for crops
13. Captive Breeding in Zoos
14. Human Cloning for Reproductive and Therapeutic Purposes
15. Genomic Editing in Humans
16. Genetically modified rice

C. **The Interplay Between Science, Society, and Sustainability**

17. Fast Fashion: The Environmental Conundrum and Societal Perspectives
18. The Use of Cannabis
19. Artificial Sweeteners

D. **Health and Nutrition Policies**

20. Fat Tax on Unhealthy Foods
21. Artificial Food Coloring
22. Food Additives

A. Workplace Dynamics and Society

1. **Balancing the Benefits and Drawbacks of Remote Work**

Since the COVID-19 pandemic, the norm has shifted significantly toward remote work from home for many employees. This shift has showcased the benefits of remote work, which have continued beyond the pandemic's guidelines, with employers actively encouraging their employees to adopt flexible working arrangements. The results were mixed; thus, people's opinions were similarly divided.

One group is in favor of working from home. Peer-reviewed studies, such as those published by the National Bureau of Economic Research (NBER), have revealed positive correlations between working from home (WFH) and increased productivity, higher profits, and reduced employee turnover, all of which are advantageous metrics for businesses. Furthermore, these studies have indicated that remote work provides employees with more personal time by eliminating the burdensome daily commutes, allowing them to allocate their time as they see fit, including more sleep and the opportunity to exercise.

However, some people think this is far from the truth. After this trend of remote or hybrid work schedules persisted for years after the pandemic, a recent study conducted by a UK-based office furnishing company raised concerns about the long-term consequences of remote work on employees. This study went viral, suggesting that remote work could lead to extreme weight gain, negative changes in posture, and overall diminished employee well-being. This portrayal, accompanied by a model depicting an unhealthy human form, strongly implies that WFH may not be as beneficial as previously thought. Given this new data, opinions on remote and hybrid work styles are now divided, and it remains uncertain whether companies should entirely abandon these approaches.

- Do you think employees should return to working exclusively in the office?

Yes No

- Explain and justify your position.
- How can you explain the difference in the conclusion reached by the scientists from the studies?
- Do you think your decision might change in the future? Why or why not? Explain.

2. **Watched on the Job**

Advances in surveillance technologies, such as biometric systems, location tracking, and employee monitoring software, have sparked debates about their effectiveness and ethical implications. These technologies, rooted in computer science and behavioral studies, are designed to improve security and productivity, but their broader impacts are still being studied.

Proponents of workplace surveillance argue that these technologies provide essential security and benefits. For instance, biometric systems, which use unique physical traits such as fingerprints or facial recognition, ensure that only authorized personnel access sensitive areas (Jain et al., 2011). Employee monitoring software can track performance metrics, helping identify inefficiencies, and boosting overall productivity. Studies in cybersecurity show that real-time surveillance reduces risks associated with data breaches and insider threats by enabling early detection of anomalies (Alneyadi et al., 2016). Furthermore, supporters point out that workplaces using transparent monitoring practices often report fewer incidents of misconduct, with evidence suggesting a direct link between surveillance and improved accountability (Martin & Freeman, 2016).

Critics, however, emphasize the ethical and psychological consequences of workplace surveillance. Research in psychology highlights how constant monitoring can lead to stress, anxiety, and diminished job satisfaction (Ball, 2010). A 2020 study on workplace behavior found that employees under extensive surveillance were 30% more likely to report feelings of distrust toward their employer (Zuboff, 2020). Additionally, opponents caution that surveillance technologies, while powerful, are not foolproof; false positives or errors in biometric systems can lead to unwarranted disciplinary actions (Bowyer, 2004). Critics also warn about the potential misuse of collected data, with legal and ethical implications for employee privacy and autonomy (Clarke, 2019).

You own a company, and productivity has been falling. Your assistant suggests implementing stricter surveillance measures.

- What do you do?
- Explain and justify your position.
- Your Chief Technology Officer disagrees with you. What does he say to convince you of his position?
- What do you say back?
- Do you think your decision might change in the future? Why or why not? Explain.

3. **Artificial Intelligence vs. Human Intelligence**

Have humans become expendable in the workforce? With the rise of artificial intelligence (AI) assistants and their role in fields like healthcare, psychological therapy, and decision-making (e.g., cancer diagnosis), it is clear that AI's use is expanding into increasingly critical areas. This has sparked debate about the effects of AI on employment. As governments and policymakers weigh their options, two competing perspectives are emerging: Will AI lead to mass unemployment, or will it create new opportunities for workers?

Supporters of artificial intelligence argue that it has the potential to boost productivity, foster creativity, and create new industries and job opportunities, particularly in the developing tech industries. They claim that AI can perform tasks more accurately and efficiently than humans, making it a significant driver of economic growth. Furthermore, they believe that by automating repetitive, menial, and administrative tasks, AI allows humans to focus on honing their talents and engaging in higher-order, creative work.

AI has proven valuable in areas like data analysis and healthcare, enabling faster, more accurate diagnoses and decision-making. For example, AI's role in diagnosing diseases and predicting treatment outcomes can lead to better care for patients. Supporters argue that AI is not intended to replace jobs but to enhance human talents by taking over tedious tasks while leaving humans to explore innovation.

The World Economic Forum (2020) predicts that by 2025, AI will lead to the creation of 97 million new jobs, which will require new skill sets and technological expertise.

On the other hand, many critics express concerns about the risks and economic consequences of AI replacing human workers. According to Frey and Osborne (2017), automation could affect as much as 47% of U.S. jobs in the next two decades, particularly targeting blue-collar industries like manufacturing, transportation, and customer service. These jobs are especially concerning because they often provide employment opportunities to individuals with lower levels of education, helping many achieve a higher standard of living. Critics fear that replacing these positions with AI will cut opportunities for upward mobility, thus entrenching poverty and widening economic inequality.

Additionally, critics suggest that as AI takes over more jobs, it will lead to alienation in the workplace, with workers feeling disconnected from their roles. This could result in diminished job satisfaction, increased stress, and mental health challenges. Another concern is that businesses employing AI will have fewer payroll costs, contributing to economic inequality as wealth becomes concentrated among employers and companies deploying AI. Critics believe that this trend risks leaving disadvantaged socioeconomic groups behind.

As AI continues to expand its role in the global economy, governments are now debating whether to slow down the development of AI to limit its risks (such as unemployment and economic disparity) or allow it to advance at its current rate, focusing instead on innovation and technological progress. These decisions have far-reaching consequences on economic policy, technological innovation, and social equity.

- **Which side of the debate do you agree with? Will AI create more opportunities, or will it lead to widespread unemployment? Why?**
- Justify your position with evidence and reasoning.
- **Imagine your friend disagrees with your position. How might they try to convince you, and how would you respond to their argument?**
- Consider both sides of the argument critically and respond thoughtfully.
- **How do scientists determine the effects of AI on the workforce? What methods or evidence are used to support their conclusions?**
- Think about the research and studies cited (e.g., *Frey and Osborne,* 2017 or *World Economic Forum,* 2020).
- **Do you think your opinion about AI's effects on employment might change in the future? Why or why not?**
- **Why do scientists sometimes reach different conclusions about AI's impact on the workforce, even when they're analyzing the same data? What could explain these differing perspectives?**

4. **TikTok, Your Time Is Up**

In recent years, we have seen the rise of TikTok. Whether it be a tool for doomscrolling, a way to add products to your wishlist, or just a forum for unwinding after a long day, it has become Gen Z's non-human companion. As screen times go up and attention spans go down, the way we consume media—and consequently, see the world—has changed. The banning of TikTok has sparked intense debate surrounding issues of security and the regulation of social media platforms at large.

Proponents of banning TikTok assert that the app poses significant risks to national security and data privacy due to its ties to the Chinese government. Concerns have been raised about the potential for data mining and surveillance by foreign entities, as well as the dissemination of propaganda and misinformation through the platform. Beyond security concerns, advocates for the ban also highlight the protection of minors from harmful content, as well as addressing addiction and mental health issues stemming from excessive social media use. Moreover, there are fears about the spread of propaganda and misinformation through the platform, with TikTok being seen as a potential tool for political manipulation and deception. By prohibiting the use of TikTok, governments aim to mitigate these risks and protect the interests of their citizens from potential exploitation by foreign powers, while also addressing broader societal concerns related to youth safety and mental health, which includes a wide scope of adverse notions, ranging from loneliness (Pop et al., 2022) to body dysmorphia (Alamayar & Khotimah, 2023; Pop et al., 2022).

On the other hand, critics of banning TikTok claim that the platform helps them find other like-minded people, discover hobbies, and learn fun facts. They point out its potential to help small businesses through the power of going viral. Additionally, due to how fast paced our quotidian life has become, most people looking to relax are not in the market for a long book or a TV show, but instead prefer a quick pick-me-up. Aside from all of that, this group claims that banning TikTok is pointless, as the younger population will move on to different social media sites. They posit that TikTok is not the reason for the commodification of showing off, as our society has been moving towards that since the 1980s. Thus, any attacks on TikTok related to consumerism (and the subsequent pollution) are seen as misplaced, as this team views TikTok as a side effect and not the problem. Finally, banning TikTok as a measure to alleviate societal problems is seen as too severe, and fails to address the core issues at play.

Your government puts the issue to a vote, and you are required to provide your opinion.

- Do you vote for the banning of TikTok?
- Explain and justify your position.
- Your friend disagrees with you. What does she say in order to convince you of her position?
- What do you say back?
- Do you think your decision might change in the future? Why or why not? Explain.

[This scenario qualifies as a Socio-Scientific Issue (SSI) because it combines scientific aspects such as concerns about data mining, social media's effects on mental health, and government surveillance with ethical, social, and societal dimensions. The debate over banning TikTok touches on issues like national security, mental health (Pop et al., 2022; Alamayar & Khotimah, 2023), and consumerism.

The opposing viewpoints represent conflicting societal priorities, illustrating the complexity of decision-making inherent in SSI.]

5. **AIm Here for You: AI in Mental Health Applications**

Artificial intelligence (AI) is increasingly being used in mental health interventions, particularly through online chatbots and digital tools. These systems aim to make mental health support more accessible, efficient, and affordable. AI mental health tools can monitor patient progress, offer mental health strategies, and provide real-time support. Despite these benefits, ethical concerns and limitations have raised debates regarding their role in therapeutic interventions.

The supporters of AI in mental health interventions believe these systems provide unique advantages. AI tools like *Woebot* and *Wysa* have demonstrated promising outcomes in helping users manage symptoms of anxiety and depression by offering consistent, on-demand mental health support (Eckelkamp, 2020). AI interventions are particularly useful for individuals in rural areas, with low socioeconomic status, or those hesitant to engage in traditional therapy (Pots et al., 2021). These tools offer structured mental health plans and allow users to explore cognitive behavioral strategies at their own pace. Additionally, AI systems are not influenced by human biases, such as appearance or tone, nor do they suffer from "off days," as human therapists might. They provide immediate support, which can be essential for real-time mental health crises (Youper, 2024).

Conversely, opponents express concerns about AI's limitations and ethical risks. One of the main issues is informed consent and patient privacy, as AI systems depend on large datasets that can be vulnerable to breaches. A lack of human oversight could put sensitive mental health information at risk (Park & West, 2023). Moreover, research shows that digital tools lack the emotional connection that comes with human interaction, a key part of successful mental health treatment (Ducharme, 2023). Chatbots often rely on pre-programmed, generic responses, which some argue fail to provide meaningful, personalized support (Petracek, 2024). Furthermore, the absence of human empathy-a vital part of therapeutic interventions—raises ethical concerns about their effectiveness as a first step toward mental health treatment.

Your state is considering introducing AI mental health chatbots as the first point of contact for individuals seeking mental health support. These chatbots would act as an initial assessment and intervention step, offering users mental health strategies and tools before referring them to in-person therapy sessions if needed. The goal is to make mental health resources more accessible, particularly for underserved or hesitant individuals. However, this proposal has sparked ethical and practical concerns, including questions about privacy, the lack of human emotional connection, and the effectiveness of AI tools. Your community is now being asked to vote on whether this approach should become the standard pathway for mental health care in your state.

- What do you vote for?
- Explain and justify your decision.
- Another expert, Professor Lara, disagrees with your position. What does she say to illustrate the reasons behind her stance and convince you?
- What would you reply to Professor Lara to explain that your position is right?
- How do scientists know about the effects of AI in mental health application?
- How certain are scientists about the knowledge related to the effects of AI in mental health applications?
- Do you think you might change your decision in the future? Explain.

6. **Gotcha: Facial Recognition and Privacy**

Facial recognition technology identifies individuals based on their facial features and has given governments and corporations significant powers in tracking individuals (Leong, 2019). While this has led to enhanced safety measures, it has also raised ethical concerns and fears of surveillance, making it a divisive issue. Public opinion is split: some view it as a critical safety tool, while others worry about misuse and violations of privacy (Leong, 2019).

Supporters of facial recognition technology believe it is a reliable way to ensure public safety and maintain national security. They highlight its efficiency in law enforcement efforts. For instance, facial recognition allows police to analyze CCTV footage to identify criminals quickly without relying solely on eyewitness accounts. This has proved especially useful at airports, where it has streamlined border control and identified overstayed visa holders. The U.S. Department of Homeland Security uses this technology to identify individuals attempting to cross borders without proper authorization (Patel, 2021).

Facial recognition has also been effective in identifying and reducing crimes. The BBC reported that the London Metropolitan Police used this technology by analyzing the 30 most problematic, anonymous shoplifting suspects from large retailers. This investigation led to the identification of 149 out of 302 suspects within just a minute (BBC News, 2023). Furthermore, the technology has even aided in locating missing persons. One notable case involved a man who was abducted as a baby and reunited with his family after being missing for 25 years thanks to facial recognition technology (Rui, 2023).

Critics argue that facial recognition comes with significant risks. They emphasize ethical concerns, particularly regarding privacy and misuse. Many believe individuals are vulnerable to being tracked without their consent, a threat that could lead to serious consequences even when technology is being employed for "public safety" purposes (Leong, 2019).

Facial recognition systems are also criticized for their inaccuracy and discrimination. A report by MIT Media Lab found that these systems are significantly less accurate when analyzing darker-skinned or female faces (Horwitz, 2018). Such biases raise concerns about wrongful accusations and convictions, particularly for marginalized groups. Some jurisdictions, such as San Francisco and Oakland, have already banned the use of facial recognition technology due to its inaccuracies and ethical implications (Metz, 2019).

Moreover, opponents are worried about a lack of oversight. The absence of strong regulation means governments and corporations could misuse these tools. A BBC report highlighted a story of a woman wrongfully accused of shoplifting, where facial recognition technology wrongly identified her, leading to severe personal distress (Clayton, 2024). Detractors argue that these risks could lead to systemic discrimination, invasion of privacy, and erosion of personal freedoms (Leong, 2019).

Your government is considering implementing facial recognition technology in all public spaces to improve security, streamline crime prevention, and make public spaces safer. This proposal would involve integrating these systems into areas such as airports, train stations, shopping districts, and other high-traffic areas. Proponents argue that this would reduce crime rates, help identify missing individuals and enhance border security. Critics, however, are raising concerns about privacy violations, inaccuracies in technology, racial bias, and the potential for government and corporate misuse. The issue has now been put to a public vote, and you and your fellow citizens must decide whether this technology should be implemented nationwide.

- What do you vote for?
- Explain and justify your decision.
- Another expert, Professor Samer, disagrees with your position. What does he say to illustrate the reasons behind his stance and convince you?
- What would you reply to Professor Samer to explain that your position is right?
- How certain are scientists about the knowledge related to the effects of implementing facial recognition software in public places?
- Do you think the knowledge about the effects of implementing facial recognition software in public places might change in the future? Explain why or why not.
- Do you think you might change your decision in the future? Explain.

B. Ethical Considerations in Scientific Research

7. The Ethical Debate Over Animal Testing in Scientific Research

Animal testing, a widespread practice involving approximately 26 million animals in the United States annually, encompasses various research purposes, including basic biology, disease study, and evaluating the efficacy and safety of medicinal and consumer goods. It involves procedures developed and tested on animals before human applications, sparking a multifaceted debate centered on scientific importance, ethical considerations, and humane alternatives.

Proponents argue that animal testing is indispensable for medical breakthroughs, citing significant contributions to treatments for conditions like breast cancer, brain injury, childhood leukemia, cystic fibrosis, and tuberculosis. Additionally, they claim that animal testing benefits the animals themselves in the grand scheme of things, as numerous animals have been saved from fatal illnesses including rabies, feline leukemia, and Potomac fever since the vaccine was initially tested on them. Moreover, supporters emphasize the necessity of animal testing in vaccine development (like with the COVID vaccine and speeding up its manufacturing) as well as its role in ethical justifications when human subjects might lead to casualties that could be easily avoided. Furthermore, animal testing has been credited with facilitating the bulk of biological discoveries (National Academies Press, 1988).

On the other end, opponents highlight the occasional harm to animal welfare in testing, with some measures like burning sometimes being included in the process. They question the very validity of animal testing, citing historical examples like Rimonabant. Rimonabant was originally developed to fight obesity and worked on mice and rats. But when the time came for human test results, it was discovered that its associated risk of depression was so high that it was banned from the market (Ettaro et al., 2020). Additionally, this group suggests that scientists use measures akin to 3D bioprinting (like organs-on-a-chip), cell culturing, and microfluidics, which is the manipulation of small fluid volumes in microscale channels—in other words, control of cellular microenvironments.

- Are you in favor of animal testing?

Yes No

- Explain your stance and justify your decision.
- Another scientist, Professor Fernando, disagrees with you. How could he explain his position and provide supporting arguments to convince you?
- How would you respond to Professor Fernando to defend your point of view?
- How do scientists know about the effects of animal testing?
- Do you think your decision might change in the future? Why or why not? Explain.
- Why did scientists reach different conclusions even though they were looking at the same data?

8. **Human Subject Research**

Clinical trials and human subjects research play a pivotal role in advancing medical knowledge, developing new treatments, and improving healthcare outcomes. These studies involve systematic investigations of potential treatments, interventions, or medical devices in human participants to evaluate their safety, efficacy, and potential side effects. However, conducting clinical trials and human subjects research raises complex ethical, legal, and scientific considerations that must be carefully navigated to ensure the protection of participants' well-being and autonomy.

Supporters of human subject research argue that these studies are vital for medical progress. They have led to groundbreaking discoveries such as the 1954 Salk Polio Vaccine Trial and the sequencing of the Human Genome Project. Supporters also emphasize that participation allows individuals with serious or life-threatening conditions to access innovative treatments that are otherwise unavailable. Clinical trials enable individuals to contribute to the collective knowledge of medicine, empowering them to shape the future of healthcare through their participation.

On the other hand, critics express concerns about the ethical implications of clinical research, particularly with psychoactive drug testing. Psychoactive drugs refer to substances that alter mood, cognition, or behavior and include antidepressants, antipsychotics, and mood stabilizers. Testing these drugs raises questions about informed consent, coercion, and the exploitation of vulnerable populations. Ethical concerns include whether participants fully understand the risks, how much pressure they face to participate, and whether the potential side effects of these drugs are adequately understood. Moreover, disparities in access to clinical trials or the exclusion of certain demographic groups can limit the generalizability of research findings and contribute to inequities in healthcare.

Activists are now planning to protest in hopes of halting human subject research for testing psychoactive drugs, citing ethical concerns such as the potential exploitation of vulnerable populations and risks to participant well-being. Your government has put the issue to a vote, and you are required to share your opinion.

- Do you vote for or against human subject research?
- Explain and justify your position.
- Your friend disagrees with you. What does she say to convince you of her position?
- What do you say back?
- Do you think your opinion might change in the future? Why or why not? Explain.
- How do ethical concerns like informed consent and autonomy influence the debate about halting psychoactive drug research?

9. **Genetically Modified Babies**

Genetically modified babies, also known as *designer babies*, involve altering an infant's genetic makeup to include specific genes or eliminate disease-associated ones. This process can occur by introducing desired genetic material into the embryo or manipulating the sperm or egg cells of parents. While some scientists support genetic modifications for disease treatment, concerns exist about potential cosmetic use and human trait enhancement, impacting broader society.

One group believes that this technology opens new horizons, with the potential to reduce the risk of certain genetic diseases, improve health outcomes, and enhance quality of life. Supporters argue that genetic engineering has been ongoing for centuries through selective breeding, and modern practices like in vitro fertilization involve sequencing embryos for disease screening. Proponents claim that these advancements could lead to improved health, reduced suffering, and even increased lifespan by decreasing the likelihood of inheriting genetic disorders.

However, opponents claim ethical issues, uncertainty about long-term effects, and concerns about the complexity of addressing diseases influenced by multiple genes. Additionally, critics fear that the use of this technology could reduce genetic diversity and create social inequalities, as access to genetic modification may not be equal for all individuals. The debate extends to societal conflicts arising from unequal access, ethical decision-making, and the possibility of prioritizing certain traits over others.

The debate continues to grow as scientists explore the effects of these genetic changes and governments weigh the scientific, ethical, and societal implications.

- Should governments support genetic modifications for their potential benefits, such as reducing genetic diseases and improving health, or should the ethical concerns, uncertainties, and risks associated with this technology lead to caution?
- How can you explain that scientists might reach different conclusions even though they are examining the same data about the effects of genetically modified babies?
- How do scientists determine the effects of genetic modifications, and what methods or evidence are they using?
- How certain are scientists about the safety, benefits, and risks associated with genetically modified babies?
- Do you think that the scientific understanding of genetically modified babies might change in the future? Why or why not?

Your government is debating whether genetically modified babies should be supported, restricted, or strictly monitored due to these ethical, scientific, and social concerns. Your input is required to make a well-reasoned decision.

10. **Genetic Testing (Applications, Controversies, and Social Perspectives)**

Genetic testing is a medical examination that identifies alterations in genes, chromosomes, or proteins. It has been available globally since the late twentieth century and offers a range of applications with varied medical and social consequences. Despite its potential benefits, genetic testing has generated diverse opinions, with some supporting its use and others opposing it.

Proponents argue that genetic testing has medical and social advantages. According to the CDC (Centers for Disease Control and Prevention), genetic testing can reveal genetic conditions within families before symptoms manifest. For example, a married woman can assess the likelihood of her child inheriting genetic diseases. Genetic testing can also uncover chromosomal abnormalities in newborns and enable early interventions. In oncology, genetic testing determines the type and progression of a patient's cancer, allowing doctors to recommend the most effective and individualized treatments. Medline Plus highlights that genetic testing can streamline health strategies by identifying diseases early, allowing for preventive monitoring and personalized therapy options.

Critics raise ethical and societal concerns about genetic testing:

1. Discrimination Risks: There is worry that individuals may face discrimination based on perceived genetic traits, particularly when social or economic disparities hinder access to genetic testing.

2. Reproductive Decisions and Societal Pressure: Opponents are concerned that genetic testing could lead to decisions, such as abortion, based on a child's genetic makeup, sparking ethical debates.
3. Anxiety and Emotional Strain: Genetic testing results can lead to heightened stress and familial strain due to concerns about accuracy and uncertainty. According to Sharon Begley's research, 57% of Americans trust genetic testing for disease prediction, while 24% are skeptical.

This debate raises ethical, social, and medical questions about whether the application of genetic testing should be encouraged or avoided.

Your city plans to apply genetic testing for conditions such as genetic disorders or risk prediction. The issue is now up for a public vote. If the majority votes in favor, the implementation will take effect next year.

- Would you vote for or against applying genetic testing in your community?
- Explain and justify your decision with clear reasoning.
- Imagine another scientist disagrees with your position. How could this scientist illustrate their arguments to convince you?
- How would you respond to that scientist to defend your original choice?
- How can you explain why scientists might come to different conclusions about the effects of genetic testing, even while reviewing the same data?
- How do scientists determine the effects and risks of genetic testing?
- How confident are scientists about their understanding of genetic testing?
- Do you think scientific knowledge about genetic testing might change in the future? Why or why not?

Additionally, how would you approach a comparison between genetic testing for genetic diseases like Lesch-Nyhan syndrome and genetic testing related to traits like musical or reading ability? Would each type of testing raise the same ethical and social concerns? How might societal and ethical perspectives differ between these applications?

Your decision and reasoning will contribute to the city's understanding of genetic testing's potential role and consequences.

11. **Modifying Legumes: Adding Vitamin D to tomatoes**

The addition of vitamin D to tomatoes in various regions of Ghana, despite the ample sunlight, has sparked controversy.

Advocates for this practice contend that Ghanaians face a vitamin D deficiency and incorporating it into a widely consumed fruit like tomatoes could effectively improve the population's vitamin D levels. They argue that it presents a safe and cost-effective means of preventing vitamin D deficiency throughout individuals' lifetimes, and many national and international scientific organizations support this supplementation.

However, opponents of adding vitamin D raise concerns about it being a form of involuntary medication. Additionally, they highlight the lack of approval from relevant health authorities, such as Ghana's Food and Drugs Authority (FDA) or similar regulatory agencies. This group also emphasizes scientific research indicating that excessive vitamin D intake-levels exceeding 4000 IU per day for adults, as recommended by the National Institutes of Health-can lead to symptoms such as weakness, dry mouth, nausea, vomiting, and potentially dangerous levels of calcium in the blood, known as hypercalcemia.

You live in Ghana and the country plans on adding vitamin D to tomatoes and requires residents to vote for or against this issue.

- Would you vote for adding vitamin D to tomatoes?

Yes No

- Explain and justify your decision.
- Another scientist, Carl Linnaeus, disagrees with your position. How could he explain his position to illustrate the reasons supporting it and convince you?
- What would you reply to Carl to explain that your position is right?
- How can you explain that scientists reached different conclusions even though scientists were all looking at the same data about the effects of adding vitamin D to tomatoes?
- Do you think the knowledge about adding vitamin D to tomatoes in Ghana might change in the future? Please explain.
- Do you think you might change your decision in the future? Please explain.

12. **Adding Fertilizers to Crops**

The issue of adding chemical substances, such as fertilizers, to crops has sparked diverse opinions among farmers and policymakers, especially regarding the potential benefits and risks associated with human health and the environment.

Proponents of adding fertilizers to crops argue that they lead to increased food production by improving crop yield and efficiency. Studies have shown that higher yields of essential crops can reduce the need for deforestation to clear new farmland, thereby supporting environmental preservation. Additionally, fertilizers can support economic growth by increasing agricultural productivity and ensuring that food supplies meet a growing population's needs.

Critics argue that the use of fertilizers in agriculture poses risks to human health and the environment. Fertilizers can lead to water contamination through runoff, which may harm ecosystems and impact human health by entering drinking water supplies. Furthermore, there are concerns that chronic exposure to certain chemical substances can lead to health risks such as developmental issues, learning disabilities, or diseases related to hormonal disruption.

Epidemiological evidence from studies suggests that prolonged exposure to agricultural runoff linked to fertilizers can impact endocrine systems and other biological pathways. Opponents also raise concerns about environmental harm, such as algal blooms caused by nutrient runoff, which can disrupt aquatic life and natural ecosystems.

Your region is debating whether to adopt increased use of fertilizers in farming practices as part of agricultural policies. If the majority votes in favor, this policy would be implemented for the next three years.

- Would you vote for or against adopting the increased use of fertilizers in agriculture?
- Explain and justify your decision.
- Another scientist, Professor Rashid, disagrees with your position. How could Professor Rashid illustrate his reasons to support increased fertilizer use and convince you?

- How would you respond to Professor Rashid to explain why your position is valid?
- How can you explain that scientists might arrive at different conclusions, even when analyzing the same data about the effects of adding fertilizers to crops?
- How do scientists determine the effects and risks of fertilizers?
- How confident are scientists about these conclusions?
- Do you think the understanding of the effects of these practices might change in the future? Explain.
- Would you consider changing your opinion in the future? Justify your reasoning.

13. **Captive Breeding in Zoos**

With the increased extinction of wildlife in our world today, many zoos are adopting the captive breeding process. Captive breeding, also known as conservation breeding, is a method employed to safeguard nearly extinct animals and plants by maintaining them in secure environments.

Those in favor of captive breeding in zoos argue that it plays a crucial role in establishing healthier generations of species not in their natural habitats. This practice is considered the optimal and final solution when a species faces the imminent threat of extinction. By preserving endangered species from extinction risks such as disease, excessive predation, and overhunting by humans, captive breeding contributes to ecosystem preservation. Zookeepers and trainers strive to build positive environments for animals, replicating their natural homes, while also ensuring better living conditions, care, medication, and protection from predators and diseases. One famous example is the captive breeding of the Wild Whooping Crane, which started in the US in 1967. Those tall white birds were at risk of going extinct due to loss of habitat, but thanks to scientists' efforts alongside government bodies, their populations have been replenished (U.S. Geological Survey, 2019).

Conversely, opponents of captive breeding highlight the significant costs associated with providing necessary resources such as food, medication, and other expenses for zoo animals. They point out that inbreeding is prevalent among captive animals, leading to issues like infertility, the expression of recessive traits, and other genetic problems. Animals born in captivity often face a lifetime of confinement. Additionally, keeping animals in captivity poses risks for caretakers, who may experience injuries from bites, stings, pounds, and other incidents. Studies indicate that confining animals in limited spaces like cages has negative physical effects, particularly for large animals like tigers and elephants. Some captive breeding programs face challenges related to inbreeding, which can lead to issues like infertility or genetic problems if not carefully managed. Critics argue that while captive breeding aims to prevent extinction, it may still raise ethical questions about animal welfare, such as the psychological and physical effects of confinement.

- Do you agree with captive breeding?

Yes No

- Explain and justify your position.

- Another researcher concerned with the world of animals disagrees with you. How would he/she develop the argument to convince you of his/her position?
- What can be your answer to this researcher who has an opposing position to yours?
- How do scientists know about the effects of captive breeding?
- Do you think your decision might change in the future? Why or why not? Explain.
- Why did scientists reach different conclusions even though they were looking at the same data?

14. **Human Cloning for Reproductive and Therapeutic Purposes**

Cloning is a controversial process that enables scientists to grow genetic duplicates of living organisms, including humans. This scenario focuses on two types of human cloning: reproductive cloning, which involves creating a genetic duplicate of a living or deceased individual, and therapeutic cloning, which involves creating embryonic stem cells to generate.

Advocates highlight the potential benefits of cloning. Reproductive cloning offers hope for infertile parents, providing a potential solution for couples struggling to conceive. Therapeutic cloning could lead to breakthroughs in regenerative medicine, such as creating tissues resistant to graft rejection and developing treatments for diseases like Parkinson's and rheumatoid arthritis.

Critics of reproductive cloning argue that it raises concerns about individuality and societal perceptions of clones, as it involves creating genetic duplicates of existing individuals. While identity is shaped by more than genetics, the societal stigma surrounding clones may impact their psychological well-being. Opponents also emphasize the ethical concerns of using embryos in therapeutic cloning and the risks of genetic defects due to the experimental nature of cloning. The trial-and-error approach may result in significant losses or disabilities in cloned organisms before achieving success.

Your city plans to use cloning for therapeutic and reproductive purposes. You are asked to vote on whether to approve these applications.

- Would you vote for using cloning for therapeutic and reproductive purposes?

Yes No

- Explain and justify your decision.
- Another scientist, Professor Pomodoro, disagrees with your position. How could he explain his position to illustrate the reasons supporting it and convince you?
- What would you reply to Professor Pomodoro to explain that your position is right?
- Do you think your decision might change in the future? Why or why not? Explain.
- Do you think the knowledge about cloning might change in the future? Explain why or why not.

15. **Genomic Editing in Humans**

The rapid advancement of biotechnology, particularly in genome editing, raises serious ethical questions with far-reaching implications. The development of CRISPR/Cas9 technology has revolutionized our ability to alter the genetic makeup of humans and other species, bringing both optimism and concern in society.

Genome editing can be categorized into somatic and germline editing. Somatic editing targets specific cells in an individual to treat diseases like sickle-cell anemia and does not affect future generations. Germline editing, however, involves modifying genes in embryos or reproductive cells, leading to changes that are passed on to offspring and raising broader ethical and societal concerns.

Proponents highlight the potential for genome editing to transform medicine by treating genetic disorders and improving human health. For example, therapeutic applications of somatic editing are already showing promise in addressing severe diseases. Skeptics, however, emphasize the ethical and scientific challenges, including risks of unforeseen side effects, mosaicism, and unintended genetic changes. Additionally, germline editing raises concerns about eugenics, socioeconomic inequality, and the possibility of "designer babies."

Organizations such as UNESCO and the World Health Organization stress the importance of informed societal debate and regulatory oversight to ensure genome editing is conducted ethically. While some advocate a cautious approach with strict limits and a focus on medical benefits, others argue for more freedom in scientific research to explore its full potential.

Despite these debates, the central challenge remains finding a balance between scientific progress and ethical responsibility to ensure the benefits of genome editing are realized without undue risk.

- Do you believe that genome editing in humans should be embraced for its potential to cure genetic diseases, or should it be approached cautiously due to ethical concerns?
- Explain and justify your position.
- Some people disagree with your position. How could they justify their position to convince you and illustrate the reasons supporting their position?
- What would you reply to these people to explain that your position is right?
- Do you think your decision might change in the future? Why or why not? Explain.
- Do you think the knowledge about genomic editing in humans might change in the future? Explain why or why not.

16. Genetically modified rice[1]

Scientists in the United Kingdom have developed a new genetically modified strain of "golden rice" to deal with vitamin A deficiency. The genetically modified rice plants contain two extra genes.

One group of scientists believe that eating the genetically modified rice with the two extra genes can help prevent blindness by improving vitamin A intake during digestion. As a result, this could help reduce childhood blindness, which affects 500,000 children worldwide each year especially in developing countries in Asia. This group argues that no studies have indicated any dangers associated with

[1] This scenario was first published in Khishfe, R. (2012). Relationship between nature of science understandings and argumentation skills: A role for counterargument and contextual factors. *Journal of Research in Science Teaching, 49*(4), 489–514. https://doi.org/https://doi.org/10.1002/tea.21012. It was later published in Khishfe, R., & Jurdak, M. (2024). Scientists' views about nature of science in the context of socioscientific. *Science & Education*. https://doi.org/10.1007/s11191-024-00600-5

genetically modified food. In 2018, the GMO golden rice seed has been marked as safe and approved by the FDA and in 2019 by Europe and New Zealand (Kelam, 2019).

Another group of scientists argue that we do not know how eating genetically modified rice (or any food) will affect us. There is no biochemical analysis of the golden rice to see how adding two genes may have changed the plant as a whole. The golden rice is still not approved in Asia. Additionally, this group is concerned that the new rice is grown in the same regions as other rice so there might be crossing over (contamination), which would change the genetic material of other rice. Once this golden rice is planted, there is no way to take it back from the environment or to control its spread (making it become a weed) and cross breed with wild rice. Unavoidable and uncontrolled cross contamination and cross breeding between the strains of rice have unpredictable outcomes that have not been tested yet and are not required to be tested before safety approval. At the same time, these scientists argue that a healthily balanced diet would be a better solution than the golden rice to deal with the vitamin A deficiency. They argue that there are methods that already work in reducing vitamin-A deficiencies such as having varied diets by eating crops that naturally grow in Asia (e.g., sweet potatoes). Some claim that poor populations have switched to 70% rice diets because other crop varieties are being exported instead of consumed. Also having a rice 'monoculture' is of economic value to Asia, with many socioeconomic and ethical issues involved (Diamond, 2019). Add to that, it has been shown that the rice will not provide the daily need of vitamin A; deficient people will still need capsule supplements. Moreover, storage of the rice even at 4 °C (39 °F) degrades the vitamin A content for up to 79%, and further degrades it to 24% when cooked.

- As a scientist, do you think the golden rice should be produced and marketed?

YES NO

- Explain and justify your decision
- Another scientist, Professor Ponso, disagrees with your decision. How could he explain his position to illustrate the reasons supporting it and convince you?
- What would you reply to Professor Ponso to explain that your decision is right?
- How can you explain that scientists reached different conclusions even though they were all looking at the same data about genetically modified rice?
- How do scientists know about the effects of genetically modified food?
- How certain are scientists about the knowledge related to genetically modified food?
- Do you think the knowledge about genetically modified food might change in the future? Explain why or why not.
- As a scientist, do you think you might change your decision in the future? Explain why or why not.

C. The Interplay Between Science, Society, and Sustainability

17. Fast Fashion

Fast fashion is the rapid production and mass consumption of inexpensive clothing items that follow the latest fashion trends. Originally, the concept of "seasons" in the fashion world was limited to four notions: "Spring/Summer," "Fall/Winter," "Resort/Cruise," and "Pre-Fall". With the advent of social media and a growing middle class, trends have replaced seasons, with some, called "microtrends", lasting only a week. This has led to the rise of fast fashion as a way for consumers to keep their wardrobes up to date. The mass manufacturing of clothing and accessories has led to an inquiry about the fast fashion industry and what goes on behind the scenes.

Proponents of fast fashion emphasize its role in the economy and its emotional value to consumers. According to the Business Research Company (2024), the global fast fashion was valued at $123 billion in 2023, and this number is only expected to grow. The industry employs 75 million people globally, many of whom are marginalized women with no formal education (Rawat, 2020). On the receiving end, consumers can look fashionable and express themselves without breaking the bank. It is believed that clothing, being constantly visible and serving as a medium of self-expression, plays a significant role in shaping one's identity (Berger & Heath, 2007). On top of that, several fast fashion companies place a high value on using plastic-free, biodegradable, and organic materials that produce no waste (Papadopoulou et al., 2022).

On the flip side, there are people that think that the fast fashion industry is a problem, citing questionable working conditions and environmental footprint. Fast fashion workers earn an average of 33 cents an hour in Bangladesh (Nguyen, 2022). The minimum wage in Bangladesh is $113, and workers can attain it through working 100+ hours per week. This has raised questions about whether governments should allow the industry to operate as is. Moving on to its environmental impact, fast fashion ranks as the second most polluting industry worldwide, following the oil industry (Vehmas et al., 2018). Approximately 87% of non-manufactured debris ends up in landfills, contributing significantly to air pollution (Moazzem et al., 2021). Moreover, the sector's high-water usage and use of harmful chemicals contribute to nearly 20% of wastewater worldwide, while synthetics and microplastics in clothing pose long-term environmental threats (Turk, 2023).

- Do you support the fast fashion industry?

Yes No

- Provide a thorough explanation for your chosen stance.
- A college student, Lily Walker, disagrees with you. What would she say to convince you of her viewpoint?
- How would you fortify your argument in response to hers?
- Have you ever purchased something from a fast fashion retailer (Shein, FashionNova, H&M, etc.)?
- Do you think all their customers are aware of the issues discussed?
- Do you think your decision might change in the future? Why or why not? Explain.

[Fast Fashion qualifies as a socio-scientific issue (SSI) because it integrates complex societal and scientific dimensions, prompting ethical and practical decision-making. The industry's reliance on advancements in material science, such as biodegradable textiles and synthetic fabrics, intersects with its significant environmental impact, including microplastic pollution, water overuse, and chemical waste. Additionally, it raises critical social concerns about exploitative labor practices, global economic reliance on marginalized workers, and consumer-driven behaviors fueled by affordability and self-expression. These interconnected aspects provoke public engagement and ethical debates, challenging individuals and policymakers to balance environmental sustainability, fair labor conditions, and societal desires for accessible fashion. By fostering informed discussions and decisions, Fast Fashion exemplifies the interplay of science, ethics, and societal impact central to SSI.]

18. **The Use of Cannabis**

Cannabis is a tall plant with a firm upstanding stem, isolated serrated leaves, and glandular hairs. It is utilized to deliver hemp fiber and as a medication. Cannabinoids are more known as medications that share dynamic specialists found in Cannabis or Marijuana. It is likewise notable in Lebanon with another natural name called, "hashish".

Even though it is progressively becoming legalized in different nations, in Lebanon it is not, since the government expects that it will hurt the populace by introducing it as another fundamental driver of addiction. In fact, they consider that pot will cause enormous harm in the public arena, particularly among teens.

Another group proposes that cannabis must be legal to use for medical and scientific causes. Based on a study, a remedy type of CBD is utilized for seizure jumble (epilepsy). CBD is additionally utilized for nervousness, torment, a muscle problem called dystonia, Parkinson sickness, Crohn infection, and numerous different circumstances, with the presence of government control. Where the government ought to apply regulations and strategies to restrict the utilization of pot just for medical behaviors. Besides, a new report tracked down that CBD and other non-psychoactive cannabinoids can successfully be utilized to forestall and treat GI issues like crabby entrail disorder (IBS), and provocative inside sickness (IBD), Crohn's, ulcerative colitis, and that's only the tip of the iceberg. A patient overview said that practically all respondents (97%) utilized clinical marijuana fundamentally for the help of persistent agony. Normal detailed help with discomfort from the clinical pot was significant. Normal pre-treatment torment on a zero to ten scale was 7.8, though normal post-treatment torment was 2.8, giving a revealed normal improvement of 5 places.

- Would you support the use of the cannabis plant therapeutically in Lebanon?

Yes No

- How do you support your claim and explain that your position is right?
- What are some effects that will occur when considering the cannabis plant legal in Lebanon?

- Do you propose any suggestions a government can do to limit the use of the cannabis plant?
- How do scientists know about the effects of using cannabis?
- Do you think your decision might change in the future? Why or why not? Explain.
- Do you think the knowledge about the use of cannabis might change in the future? Explain why or why not.

19. **Sweet Debate: The Use of Artificial Sweeteners**

Artificial sweeteners are synthetic compounds manufactured in laboratories or derived from natural sources (Hopkins Medicine, n.d.). They are widely found in sugar-free, keto-friendly, and diet-labeled products and have become a common sugar substitute in modern diets. However, the debate over their health benefits and risks continues to divide scientific opinion and public perception.

Proponents of artificial sweeteners argue that they provide several health advantages, especially for individuals trying to reduce calorie consumption or manage certain health conditions. One of the most notable benefits is that they are low in calories or contain no calories at all, making them a popular choice for those aiming to lower their overall daily calorie intake (Hopkins Medicine, n.d.).

While some studies suggest a potential connection between artificial sweeteners and heart disease or mortality, it is emphasized that other dietary habits or behaviors can have a far greater impact (Mayo Clinic, 2023).

Another key benefit is the prevention of tooth decay. Unlike sugar, which ferments bacteria in the mouth and leads to the production of harmful acids, artificial sweeteners do not have this effect. This means they can help reduce the risk of cavities and promote better oral health (Altoona Smiles, 2023).

Some sweeteners are derived from natural sources, such as *Stevia*, which is a herb and considered safe with several health benefits. A 2017 study highlighted that *Stevia* can help manage conditions like diabetes, obesity, high blood pressure, inflammation, and kidney problems (Cleveland Clinic, 2024).

On the other side of the debate, critics point out that artificial sweeteners may bring their own set of health risks and problems. Melissa Young, a medical expert, warns that artificial sweeteners can be more harmful than natural sugar. She emphasizes that they are far sweeter than natural sugar-up to 700 times sweeter-and may overstimulate the brain's dopamine release, leading to increased sugar cravings and overconsumption (Cleveland Clinic, 2023).

Additionally, studies suggest that daily consumption of artificially sweetened beverages may raise the risk of developing type 2 diabetes by as much as 67% (McCoy, 2023).

Another major concern is the impact on the digestive system. Artificial sweeteners are associated with *dysbiosis*, a condition that leads to an imbalance in gut bacteria. This imbalance has been linked to symptoms such as bloating, migraines, mood disturbances, and other health issues (Foley, 2023).

Critics argue that instead of replacing sugar with artificial sweeteners, the best approach is to reduce sugar consumption overall, adopting moderation rather than substituting one harmful food for another.

Your government is considering implementing a nationwide policy to completely substitute all sugar with artificial sweeteners across all food products. This would involve removing traditional natural sugar from the food supply entirely, replacing it with artificial sweeteners in all sectors, including processed foods, beverages, and other sugar-dependent items. Proponents believe this shift will lead to health improvements, lower calorie intake, and better management of diabetes and obesity rates. However, critics worry about health risks associated with artificial sweeteners, the potential for increased sugar cravings, effects on the gut microbiome, and the risk of developing chronic diseases like type 2 diabetes as a result of consistent exposure to artificial sweeteners.

The issue has now been put to a public vote, and you must decide whether this major change should be enacted nationwide.

(a) What do you vote for?
(b) Explain and justify your decision.
(c) Another expert, Professor Lana, disagrees with your position. What does she say to illustrate the reasons behind her stance and convince you?
(d) What would you reply to Professor Lana to explain that your position is right?
(e) How do scientists know about the effects of artificial sweeteners?
(f) How certain are scientists about the knowledge related to artificial sweeteners?
(g) Do you think the knowledge about artificial sweeteners might change in the future? Explain why or why not.

D. Health and Nutrition Policies

20. Fat Tax on Unhealthy Foods

Fat taxes, also referred to as junk food taxes or sugary beverages taxes, are additional charges or fees applied to foods and beverages deemed unhealthy. These taxes primarily aim to reduce consumption rates of items with high levels of fat, sugar, and salt, which are directly associated with health risks such as diabetes, obesity, and cardiovascular conditions. The goal is to encourage the consumption of healthier options and habits.

There are diverse and varied viewpoints on the effectiveness and utility of fat taxes. Some stakeholders believe in their high positive impact, while others hold the opposite view.

Supporters of fat taxes on unhealthy foods argue that implementing these taxes encourages consumers to choose healthier food and beverage options, limiting their excessive spending on unhealthy ones. These groups include Public Health Organizations and advocates, such as the World Health Organization (WHO), the American Heart Association (AHA), and the Centers for Disease Control and Prevention (CDC), who endorse the application of taxes on fatty foods as a strategy to address diet-related health issues. Health care professionals and researchers also validate this approach, believing in its positive impact on public health outcomes. Physicians, nurses, doctors, and researchers see these levies as a successful strategy for reducing dietary intake of foods and drinks high in fat, sugar, or salt, which are linked to chronic illnesses like obesity, diabetes, and heart disease. Additionally, some economists and policy experts argue that fat taxes can help address market failures related to the consumption of unhealthy foods by reflecting the social costs of these items, promoting better options, and generating income that can be used to fund public health programs.

On the other hand, various groups oppose fat taxes for a variety of reasons. Low-income advocacy groups, representing low-income individuals, express concerns about the regressive nature of these fat taxes. Research supports this view, with findings indicating that "Critics say a flat tax would hit the poor hardest because they spend 30 per cent of their income on food." Libertarian organizations and individuals argue that consumers have the freedom and the right to purchase and consume their food choices without government rules or taxes. Food and beverage industries, along with free market advocates and some economists, defend their position by stating that such taxes may be unfair, as they target specific products or industries without achieving the intended goal of changing consumer behavior. They also argue that there may be alternative approaches to addressing health issues in society.

- Would you vote for a fat tax if your country plans to implement this decision?

Yes No

- Explain and justify your answer.
- Do you think both viewpoints, in favor and against, have sufficient reasoning, arguments, and research to support their positions? How?
- Do you believe it is ethical or unethical to implement such a strategy on the public to attempt to limit obesity? Why?
- Another party strongly disagrees with your opinion. What could be their reasoning and justification?
- How would you defend your position, and what arguments would you use to try and convince the opposing party? Explain.
- Do you think your decision might change in the future? Why or why not? Explain.

21. **Artificial Food Coloring**

Artificial food coloring is an often-used element in the production of candy, beverages, and other types of junk food. It is used as a marketing tactic to give food the appearance of being fresh or a certain flavor (pink for strawberry, yellow for banana, and so on). The profits associated with this eye-catching design choice speak for themselves, and artificial food coloring is consumed by nearly everyone in first-world society, from children to adults. But the chemicals it is produced from, synthetic and often petroleum-based, have sparked controversy (Bailey & Crezo, 2022).

Advocates argue that synthetically added food colorings enhance the appearance and color of the foods, compensating for potential sensory and color loss during the phases of processing and storage. Adding food dyes is believed to preserve color, flavor, and even vitamins, which may otherwise be compromised by factors like light, temperature, and air. Additionally, the Food and Drug Administration (FDA) approval under specific conditions is cited as evidence of the safety of consuming artificially colored foods (U.S. Food & Drug Administration, 2023).

Researchers with alternative viewpoints highlight associations between the consumption of artificially colored foods and various health issues. Concerns include neurobehavioral effects, such as hyperactivity and distraction in children (Seltenrich, 2021). The American National Library of Medicine indicates immune reactive disorders associated with increased consumption of artificially colored foods (Vojdani & Vojdani, 2015). It is no surprise that even laypeople have drawn the link between cancer and food coloring. Environmental concerns are also raised, with studies asserting that many artificial dyes are not eco-friendly, posing detrimental effects on the environment by accumulating in rivers and on land (Affat, 2021).

You work as a food scientist at a food production company. Your boss just ordered the production of a new snack, and you are considering whether to add artificial food coloring or not. You walk into a meeting with your team members to announce your decision.

- What is your decision?
- How would you explain your decision to your team?
- Your project-mate Ed Ebbel, a fellow food scientist, disagrees with you. What does he say to convince you and the team members otherwise?
- How do you respond to Ed Ebbel?
- How do scientists know about the effects of artificial food coloring?
- Do you think your decision might change in the future? Why or why not? Explain.
- Why did scientists reach different conclusions even though they were looking at the same data?

22. **Food Additives**

According to the World Health Organization, food additives are substances incorporated into processed foods for technical purposes, such as extending the amount of time a food can be stored, improving safety, or modifying the sensory properties of food such as the sight, smell, taste, and touch of food (2024). Food additives can be either chemically synthesized or derived from plants, animals, or minerals. Each additive is used for a different purpose and there are several thousands of them (World Health Organization, 2023).

Many people support incorporating food additives as they serve multiple purposes, including extending the shelf life of food ensuring safety and freshness

(Lalani et al., 2024), as well as enhancing taste and sensory appeal. This group also challenges the idea that just because something is done in the lab, it must be unhealthy. Certain food additives are endorsed by expert dieticians for weight loss, as they are believed to aid in burning excess body fat without adding extra calories (Reeuwijk et al., 2014). These additives are claimed to act as fat blockers, recommended to individuals aiming to lose weight (Reeuwijk et al., 2014). Another example is calcium propionate, which inhibits the growth of mold and bacteria in bread and other baked goods. Similarly, other additives with valuable functions are emulsifiers like mono- and di-glycerides, which prevent water and oil from separating in products like mayonnaise and salad dressings (Vandergriendt, 2023).

Those against additives, on the other hand, decry their effects on health. Studies have shown that certain additives such as synthetic colorings and preservatives might contribute to hyperactivity in children (Nigg et al., 2012). Additionally, worries have sprung up concerning potential allergic reactions brought on by the chemical concoctions in processed foods because of additives. On top of that, food additives like aspartame (used in chewing gum and ice cream), saccharin (a calorie-free sweetener), and benzoates (used in sauces and tomato paste) have been shown to cause cancer in animals (Fairweather & Swann, 1981). Not to mention, the production of these additives could harm the environment. Even in the production of natural food additives, the runoff associated with growing the necessary crops may contribute to water pollution, potentially impacting aquatic systems (Carpenter & Bennett, 2011).

- Which side do you agree with?
- Explain and justify your position.
- Ed Ebbel disagrees with you. What does he say to convince you otherwise?
- How do you defend your position?
- How can you explain that scientists reached different conclusions even though scientists were all looking at the same data about the effects of food additives?
- How do scientists know about the effects of food additives?
- How certain are scientists about the knowledge related to food additives?
- Do you think the knowledge about food additives might change in the future? Explain why or why not.

References

Affat, S. S. (2021, May). *Classifications, advantages, disadvantages, toxicity effects of natural and synthetic dyes: A review*. Retrieved from https://www.researchgate.net/publication/351885086_Classifications_Advantages_Disadvantages_Toxicity_Effects_of_Natural_and_Synthetic_Dyes_A_review

Alamayar, I. H., & Khotimah, K. (2023). The impact of TikTok on body image: A narrative review of the literature. *Jurnal Ilmu Komunikasi dan Media Sosial, 3*(3), 764–773. https://doi.org/10.47233/jkomdis.v3i3.1265

Alneyadi, S., Sithirasenan, E., & Muthukkumarasamy, V. (2016). A survey on data leakage prevention systems. *Journal of Network and Computer Applications, 62*, 137–152. https://doi.org/10.1016/j.jnca.2016.01.008

Altoona Smiles. (2023, November 21). *Are artificial sweeteners bad for your teeth?* Retrieved from https://altoonasmiles.com/blog/are-artificial-sweeteners-bad-for-your-teeth/#:~:text=In%20summary%2C%20artificial%20sweeteners%20are,to%20ensure%20optimal%20dental%20health

Bailey, M. M., & Crezo, A. (2022, November 4). *Artificial colorings (synthetic food dyes).* Center for Science in the Public Interest. Retrieved April 9, 2024 from https://www.cspinet.org/article/artificial-colorings-synthetic-food-dyes

Ball, K. (2010). Workplace surveillance: An overview. *Labor History, 51*(1), 87–106. https://doi.org/10.1080/00236561003654776

BBC News. (2023). Facial recognition technology helps identify shoplifting suspects. https://www.bbc.com/news/technology-65027344

Berger, J., & Heath, C. (2007). Where consumers diverge from others: Identity-signaling and product domains. *Journal of Consumer Research, 34*(2), 121–134. https://doi.org/10.1086/519142

Bowyer, K. W. (2004). Face recognition technology: Security versus privacy. *IEEE Technology and Society Magazine, 23*(1), 9–19. https://doi.org/10.1109/MTAS.2004.1273467

Carpenter, S. R., & Bennett, E. M. (2011). Reconsideration for the planetary boundary of phosphorus. *Environmental Research Letters, 6*(1). https://doi.org/10.1088/1748-9326/6/1/014009

Clarke, R. (2019). The regulation of technologies that have transboundary impacts. *Computer Law & Security Review, 35*(5), 105318. https://doi.org/10.1016/j.clsr.2019.105318

Clayton, J. (2024, May 26). *I was misidentified as shoplifter by facial recognition tech.* British Broadcasting Corporation. Retrieved from https://www.bbc.com/news/technology-69055945

Cleveland Clinic. (2023, April 20). *This is why artificial sweeteners are bad for you.* Retrieved from https://health.clevelandclinic.org/whats-worse-sugar-or-artificial-sweetener

Cleveland Clinic. (2024, June 28). *Stevia is sweet – But is it good for y*ou? Retrieved from https://health.clevelandclinic.org/what-is-stevia

Diamond, C. (2019). Golden rice and apples sliced: Staging GMO controversy in *Snow White and the Apple's Revenge. Research in Drama Education, 24*(4), 522–528. https://doi.org/10.1080/13569783.2019.1643711

Ducharme, J. (2023, October 4). *Can AI chatbots ever replace human therapists?* Time. Retrieved from https://time.com/6320378/ai-therapy-chatbots

Eckelkamp, S. (2020, December 9). *The totally free way to deal with anxiety.* Cleanplates. Retrieved from https://cleanplates.com/wellness/mental-health-bots-for-anxiety/

Ettaro, R., Laudermilk, L., Clark, S. D., & Maitra, R. (2020). Behavioral assessment of rimonabant under acute and chronic conditions. *Behavioural Brain Research, 390*, 112697. https://doi.org/10.1016/j.bbr.2020.112697

Fairweather, F. A., & Swann, C. A. (1981). Food additives and cancer. *Proceedings of the Nutrition Society, 40*(1), 21–30. https://doi.org/10.1079/PNS19810006

Foley, J. (2023, July 20). *What are artificial sweeteners, and are they bad for me? 6 possible risks to consider.* GoodRx. Retrieved from https://www.goodrx.com/well-being/diet-nutrition/artificial-sweeteners

Frey, C. B., & Osborne, M. A. (2017). The future of employment: How susceptible are jobs to computerisation? *Technological Forecasting and Social Change, 114*, 254–280. https://doi.org/10.1016/j.techfore.2016.08.019

Hopkins Medicine. (n.d.). *Facts about sugar and sugar substitutes.* Retrieved July 15, 2024, from https://www.hopkinsmedicine.org/health/wellness-and-prevention/facts-about-sugar-and-sugar-substitutes

Horwitz, J. (2018, February 13). *If you're a darker-skinned woman, this is how often facial recognition software decides you're a man.* Quartz. Retrieved from https://qz.com/1205604/if-youre-a-darker-skinned-woman-the-best-facial-recognition-software-frequently-thinks-youre-a-man

Jain, A. K., Flynn, P., & Ross, A. A. (2011). *Handbook of biometrics.* Springer.

Kelam, I. (2019). Holistic approach to the environment and GMOs – The case of golden rice. *Global Bioethics Enquiry, 7*(3), 118. https://doi.org/10.38020/GBE.7.3.2019.118-127

Lalani, A. R., Pouyani, N. R., Askari, A., Tavajohi, S., Akbari, S., & Jafarzadeh, E. (2024). Food additives, benefits, and side effects: A review article. *Journal of Chemical Health Risks, 14*(1), 1–10. https://doi.org/10.22034/jchr.2023.1967340.1619

Leong, B. (2019). Facial recognition and the future of privacy: I always feel like… somebody's watching me. *Bulletin of the Atomic Scientists, 75*(3), 105–115. https://doi.org/10.1080/00963402.2019.1604886

Martin, K., & Freeman, R. E. (2016). Some problems with employee monitoring. *Journal of Business Ethics, 139*(3), 491–503. https://doi.org/10.1007/s10551-015-2652-5

Mayo Clinic. (2023, January 10). *Artificial sweeteners and other sugar substitutes*. Retrieved from https://www.mayoclinic.org/healthy-lifestyle/nutrition-and-healthy-eating/in-depth/artificial-sweeteners/art-20046936#:~:text=If%20you%20replace%20added%20sugar,weight%20in%20the%20short%20term

McCoy, K. (2023, November 14). *5 worst artificial sweeteners, plus healthy alternatives*. Dr. Axe. Retrieved from https://draxe.com/nutrition/artificial-sweeteners/

Metz, R. (2019, July 17). *Beyond San Francisco, more cities are saying no to facial recognition*. Cable News Network. Retrieved from https://edition.cnn.com/2019/07/17/tech/cities-ban-facial-recognition/index.html

Moazzem, S., Wang, L., Daver, F., & Crossin, E. (2021). Environmental impact of discarded apparel landfilling and recycling. *Resources, Conservation and Recycling, 166*, 105338. https://doi.org/10.1016/j.resconrec.2020.105338

Nguyen, L. (2022, October 10). *Fast fashion: The danger of sweatshops*. Earth.Org. Retrieved from https://earth.org/sweatshops/

Nigg, J. T., Lewis, K., Edinger, T., & Falk, M. (2012). Meta-analysis of attention-deficit/hyperactivity disorder or attention-deficit/hyperactivity disorder symptoms, restriction diet, and synthetic food color additives. *Journal of the American Academy of Child & Adolescent Psychiatry, 51*(1), 86–97. https://doi.org/10.1016/j.jaac.2011.10.015

Papadopoulou, M., Papasolomou, I., & Thrassou, A. (2022). Exploring the level of sustainability awareness among consumers within the fast-fashion clothing industry: A dual business and consumer perspective. *Competitiveness Review, 32*(3), 350–357. https://doi.org/10.1108/CR-04-2021-0061

Park, E., & West, D. M. (2023, November 30). *Why mental health apps need to take privacy more seriously*. Brookings. Retrieved from https://www.brookings.edu/articles/why-mental-health-apps-need-to-take-privacy-more-seriously/

Patel, G. (2021, April 11). *Considering the use of facial recognition technology in higher education*. Triangle Privacy Hub. Retrieved from https://www.triangleprivacyhub.org/wp-content/uploads/2021/04/FRinHigherEdFinal.pdf

Petracek, L. J. (2024, July 17). AI chatbots for mental health: Opportunities and limitations. *Psychology Today*. Retrieved from https://www.psychologytoday.com/us/blog/the-psyche-pulse/202407/ai-chatbots-for-mental-health-opportunities-and-limitations

Pop, L. M., Iorga, M., & Iurcov, R. (2022). Body-esteem, self-esteem, and loneliness among social media young users. *International Journal of Environmental Research and Public Health, 19*(9), 5064.

Pots, C., Ennis, E., Bond, R., Mulvenna, M., Mctear, M., Boyd, K., Broderick, T., Malcolm, M., Kuosmanen, L., Nieminen, H., Vartiainen, A., Kostenius, C., Cahill, B., Vakaloudis, A., McConvey, G., & O'Neill, S. (2021). Chatbots to support mental wellbeing of people living in rural areas: Can user groups contribute to co-design? *Journal of Technology in Behavioral Science, 6*, 652–665. https://doi.org/10.1007/s41347-021-00222-6

Rawat, A. (2020, October 30). *Fast fashion isn't saving us. Far from it.* Let Me Breathe. Retrieved from https://letmebreathe.in/2020/10/30/fast-fashion-isnt-saving-us-far-from-it/

Reeuwijk, N. M., Venhuis, B. J., de Kaste, D., Hoogenboom, R. L., Rietjens, I. M., & Martena, M. J. (2014). Active pharmaceutical ingredients detected in herbal food supplements for weight loss sampled on the Dutch market. *Food Additives & Contaminants: Part A, 31*(11), 1783–1793. https://doi.org/10.1080/19440049.2014.958574

Rui, D. (2023, December 4). *In Hebei, AI tech reunites abducted son with family after 25 years*. Sixth Tone. Retrieved from https://www.sixthtone.com/news/1014206

Seltenrich, N. (2021, May 3). *Food dyes linked to attention and activity problems in children*. Environmental Health News. Retrieved from https://www.ehn.org/food-dyes-children-health-2652857895.html

The Business Research Company. (2024). *Fast fashion global market report*. Retrieved from https://www.thebusinessresearchcompany.com/report/fast-fashion-global-market-report#:~:text=The%20fast%20fashion%20market%20size,(CAGR)%20of%20 15.5%25

Turk, V. (2023, October 2). *Fast fashion: Polluting our water millions of clothing items at a time*. George Washington University. Retrieved from https://blogs.gwu.edu/law-gwpointsource/2023/10/02/fast-fashion-polluting-our-water-millions-of-clothing-items-at-a-time/#:~:text=The%20fashion%20industry%20is%20the,producing%2020%25%20of%20 wastewater%20worldwide

U.S. Food and Drug Administration. (2023, December 14). *Color additives questions and answers for consumers*. Retrieved from https://www.fda.gov/food/color-additives-information-consumers/color-additives-questions-and-answers-consumers

U.S. Geological Survey. (2019, March 13). *Significant milestone in whooping crane recovery*. Retrieved from https://www.usgs.gov/news/featured-story/significant-milestone-whooping-crane-recovery#:~:text=The%20captive%20breeding%20program%20began,to%20the%20 USGS%20in%201996

U.S. National Research Council. (1988). *Benefits derived from the use of animals*. Use of Laboratory Animals in Biomedical and Behavioral Research - NCBI Bookshelf. National Academies Press. Retrieved from https://www.ncbi.nlm.nih.gov/books/NBK218274/

Vandergriendt, C. (2023, December 22). *What are monoglycerides and are they safe to consume?* Healthline. Retrieved from https://www.healthline.com/health/food-nutrition/monoglycerides#trans-fat

Vehmas, K., Raudaskoski, A., Heikkilä, P., Harlin, A., & Mensonen, A. (2018). Consumer attitudes and communication in circular fashion. *Journal of Fashion Marketing & Management, 22*(16), 286–300. https://doi.org/10.1108/JFMM-08-2017-0079

Vojdani, A., & Vojdani, C. (2015). Immune reactivity to food coloring. *Alternative Therapies in Health and Medicine, 21*(Suppl 1), 52–62. Retrieved from https://pubmed.ncbi.nlm.nih.gov/25599186/#:~:text=The%20molecules%20of%20synthetic%20colorants,and%20 disrupt%20the%20immune%20system

World Economic Forum. (2020, October 20). *The future of jobs report 2020*. Retrieved from https://www.weforum.org/reports/the-future-of-jobs-report-2020

World Health Organization. (2023, November 16). *Food additives*. Retrieved from https://www.who.int/news-room/fact-sheets/detail/food-additives

Youper. (2024). *Introduction page*. Retrieved July 18, 2024 from https://www.youper.ai/

Zuboff, S. (2020). *The age of surveillance capitalism: The fight for a human future at the new frontier of power*. PublicAffairs.

Chapter 8
Implications and Recommendations

Abstract This chapter examines the implications and recommendations for incorporating SSI into science education, focusing on its role in promoting scientific literacy, critical thinking, and informed decision-making. SSI-based learning connects theoretical science to practical applications, enabling students to engage in debate and evidence-based reasoning. According to research, organized reasoning is important for strengthening students' abilities to appraise arguments, negotiate complicated scientific quandaries, and engage in democratic debate. Additionally, the chapter emphasizes the need of teacher training and multidisciplinary collaboration in successfully implementing SSI curriculum. To increase students' involvement with science and society, educators must be prepared to use educational tools that combine argumentation, ethical issues, and global views. In addition, policy proposals include professional development programs, curricular improvements, and ongoing research on SSI's efficacy. By incorporating SSI into science education, educators may develop scientifically literate individuals capable of tackling current social concerns.

Keywords Scientific literacy · Argumentation · Critical thinking · Curriculum development · Teacher training

The integration of SSI into science classrooms holds profound implications for educational practices. SSI is a revolutionary method in science education that bridges the gap between abstract scientific principles and their practical applications (Sadler et al., 2007). Adopting SSI as a teaching approach allows educators to contextualize science learning within socially meaningful frameworks, making it more relevant and engaging for students. This approach is especially important in a fast-changing world, where technical and socioeconomic issues require individuals who can think critically, assess evidence, and make informed decisions. Presley et al. (2013) emphasize the importance of an educational framework for SSI-based scientific instruction, which includes design components, learner experiences, and instructor attributes that are influenced by classroom, school, and community settings. Such a framework connects scientific literacy to social, political, and ethical perspectives,

R. Khishfe, *The Complexities of Socioscientific Issues in Education*, Springer Texts in Education, https://doi.org/10.1007/978-3-031-88006-3_8

highlighting the importance of SSI-based education in encouraging critical and reflective thinking.

Recent research underscores the necessity of teaching argumentation as a discourse practice in science education, as it enhances understanding by fostering explanations, models, and theories while engaging students in social dialogues with opposing viewpoints (Erduran & Jiménez-Aleixandre, 2008). Within the SSI framework, argumentation equips students to critically assess claims and navigate controversial, transdisciplinary topics, promoting critical thinking and active democratic participation. This approach enriches students' understanding of scientific epistemology, making argumentation a cornerstone of effective SSI-based instruction.

The positive relationship observed between arguing and the understanding of NOS aspects within the SSI framework, as demonstrated in studies like Khishfe's (2023), suggests that incorporating SSI can enhance students' argumentation skills and deepen their understanding of the epistemology of science. This connection implies that argumentation serves as a critical tool for engaging students with NOS, fostering their ability to critically evaluate scientific evidence and its societal applications. By integrating argumentation into SSI-based learning, educators can bridge the gap between theoretical knowledge and practical reasoning, enabling students to navigate complex and controversial scientific issues. Moreover, this approach aligns with the broader goal of developing scientifically literate citizens who can participate in informed decision-making processes in a rapidly evolving world. To fully realize these benefits, educators must implement structured opportunities for argumentation within SSI contexts, emphasizing not only evidence-based reasoning but also the ethical, social, and cultural dimensions of science. This dual emphasis prepares students to engage with multifaceted global challenges, highlighting the critical role of SSI in cultivating a deeper and more nuanced understanding of science and its impact on society.

Topçu et al. (2014) identify teacher preparation as a critical factor in the successful integration of SSI into science education. Pre-service teachers need targeted training to build self-efficacy, informal reasoning skills, and the ability to address complex societal issues through SSI-based instruction. Embedding SSI-focused modules in teacher education programs is essential to overcoming challenges such as limited training and inadequate resources. Macalalag et al. (2020) further highlight the transformative impact of interdisciplinary and culturally diverse experiences, such as those in Pennsylvania, USA, and Sicily, Italy, which broaden teachers' perspectives on SSI. These experiences underscore the need for interdisciplinary collaboration and the integration of technology to design effective SSI curricula. By linking scientific concepts to real-world issues, these strategies not only prepare students to tackle societal challenges but also emphasize the broader implications of equipping educators to lead meaningful SSI-based instruction.

In educational contexts, distinguishing between arguing and argumentation is key for understanding how students acquire critical thinking and reasoning abilities. Arguing is sometimes described as an informal, emotive debate in which people want to "win" by convincing others of their point of view, regardless of evidence or logical consistency. In contrast, argumentation is an organized, evidence-based process that promotes collaborative problem-solving and critical thinking. According to Osborne (2010), argumentation is essential for scientific literacy because it helps students engage with evidence, develop explanations, and grasp scientific epistemology. This is consistent with Presley et al.'s (2013) Vision II of scientific literacy, which emphasizes real-world science applications such as data analysis and reasoning in socially and morally important situations. Socioscientific questions are particularly effective in building Vision II literacy because they demand students to weigh scientific processes against ethical, political, and social factors. SSI integration should also prioritize cultural and global relevance, with scientific themes contextualized within local and international contexts. For example, addressing climate change through SSI in a global perspective may emphasize the disparities in effects between countries while motivating students to seek equitable and culturally appropriate solutions.

According to Sadler (2011), a developing paradigm for SSI-based education emphasizes the need to engage students in higher-order SSI negotiation processes such as argumentation, decision-making, and position-taking. Additionally, students should be able to analyze scientific data, debate associated scientific hypotheses, and investigate the political and economic implications of SSI themes. Similarly, Sadler (2004) emphasizes that reasoning in the context of SSI problems challenges students to balance scientific evidence alongside ethical, social, and cultural factors, encouraging deeper engagement with scientific concepts. Although ethical considerations are valuable in SSI education, Sadler (2011) suggests that exploring broader social and scientific dimensions is equally effective in enhancing critical thinking. Furthermore, systematic argumentation, as highlighted by Erduran and Jiménez-Aleixandre (2008), helps students acquire metacognitive abilities by reflecting on their reasoning processes and evaluating the validity of evidence. Kuhn (1991) further contrasts daily conflicts with educational argumentation, noting that the latter fosters higher-order thinking abilities like assessing conflicting assertions and making evidence-based conclusions.

Building on these concepts, Kuhn (1993) advocates merging scientific and argumentative thinking by actively involving students in reasoning processes across various academic fields, particularly those addressing socioscientific themes. For instance, students can develop and refine their ideas on such topics through an argumentation framework that includes exploring alternative hypotheses, counterarguments, and rebuttals. This approach not only enhances scientific reasoning but also equips students with transferable skills applicable across disciplines. Presley et al. (2013) highlight the importance of fostering a supportive and collaborative classroom environment, alongside scaffolding strategies, to facilitate higher-order thinking and effective engagement with SSI. By integrating structured argumentation practices, educators can align SSI-based instruction with modern educational goals, such as fostering teamwork, ethical reasoning, and civic participation (Sadler et al., 2007; Simon et al., 2006). These implications underscore the transformative potential of SSI in preparing students for the complexities of the twenty-first century.

Therefore, educators should consider including SSI scenarios in their classrooms to improve not only scientific content knowledge but also the development of scientific reasoning, critical thinking, and decision-making capabilities. The role of argumentation within SSI scenarios further emphasizes its potential for developing students' ability to engage in evidence-based discussions and address complex issues that combine scientific and social dimensions. Secondly, the results from various studies show the potential of SSI to address the challenges of a rapidly changing modern world and, consequently, prepare students to make informed decisions as citizens. Overall, the implications suggest that SSI can serve as a tool that provides a more engaging science education that empowers students with the skills they need to actively participate in societal discourses. Moreover, according to research, SSI is an excellent platform for improving argumentation abilities since students are encouraged to draw on scientific evidence while also considering ethical and social dimensions (Sadler, 2004). For example, students discussing the safety and ethics of nuclear energy must weigh evidence on energy efficiency against concerns about environmental dangers and waste disposal, an activity that enhances both their scientific reasoning and public discourse skills.

Finally, integrating SSI into science curricula, as supported by research-based frameworks such as Presley et al.'s (2013), underscores the importance of SSI in building scientific literacy and preparing students to address social and scientific dilemmas. Building on the identified implications, several recommendations arise for educators, curriculum developers, and policymakers. Firstly, there is a need for professional development programs that help educators gain the skills needed to effectively implement SSI into their teaching practices. To assist educators, schools should invest in long-term professional development activities such as seminars, mentorship programs, and online learning platforms. Such activities can assist instructors in acquiring confidence in SSI implementation while also staying current on emerging scientific and societal concerns, ensuring that their education remains relevant and meaningful. Policymakers should provide clear standards and funding mechanisms for SSI implementation at the national and regional levels. This involves incorporating SSI into teacher certification standards and investing resources to create culturally responsive educational materials that address both local and global scientific challenges. Moreover, curriculum developers should collaborate with educators to develop SSI-based instructional materials that cater to diverse learning needs and cultural contexts. Secondly, policymakers can advocate for SSI to be incorporated by educational institutions as a core component of a science curriculum and integrate it across grade levels. Lastly, ongoing research is recommended to continually assess the effectiveness of SSI integration in the development of content and skills. By following these recommendations, educators and policymakers can collectively contribute to the realization of the full potential of SSI in shaping a more engaging science classroom.

References

Erduran, S., & Jiménez-Aleixandre, M. P. (Eds.). (2008). *Argumentation in science education: Perspectives from classroom-based research* (1st ed.). Springer. https://doi.org/10.1007/978-1-4020-6670-2

Khishfe, R. (2023). Relationship between nature of science and argumentation: A follow-up study. *International Journal of Science and Mathematics Education, 21*(4), 1081–1102. https://doi.org/10.1007/s10763-022-10307-0

Kuhn, D. (1991). *The skills of argument*. Cambridge University Press. https://doi.org/10.1017/cbo9780511571350

Kuhn, D. (1993). Science as argument: Implications for teaching and learning scientific thinking. *Science Education, 77*(3), 319–337. https://doi.org/10.1002/sce.3730770306

Macalalag, A. Z., Johnson, J., & Lai, M. (2020). How do we do this: Learning how to teach socioscientific issues. *Cultural Studies of Science Education, 15*(2), 389–413. https://doi.org/10.1007/s11422-019-09944-9

Osborne, J. (2010). Arguing to learn in science: The role of collaborative, critical discourse. *Science (American Association for the Advancement of Science), 328*(5977), 463–466. https://doi.org/10.1126/science.1183944

Presley, M. L., Sickel, A. J., Muslu, N., Merle-Johnson, D., Witzig, S. B., Izci, K., & Sadler, T. D. (2013). A framework for socio-scientific issues-based education. *Science Educator, 22*(1), 26.

Sadler, T. D. (2004). Informal reasoning regarding socioscientific issues: A critical review of research. *Journal of Research in Science Teaching, 41*(5), 513–536. https://doi.org/10.1002/tea.20009

Sadler, T. D. (2011). Socio-scientific issues-based education: What we know about science education in the context of SSI. In T. D. Sadler (Ed.), *Socio-scientific issues in the classroom* (pp. 355–369). Springer. https://doi.org/10.1007/978-94-007-1159-4_20

Sadler, T. D., Barab, S. A., & Scott, B. (2007). What do students gain by engaging in socioscientific inquiry? *Research in Science Education (Australasian Science Education Research Association), 37*(4), 371–391. https://doi.org/10.1007/s11165-006-9030-9

Simon, S., Erduran, S., & Osborne, J. (2006). Learning to teach argumentation: Research and development in the science classroom. *International Journal of Science Education, 28*(2–3), 235–260. https://doi.org/10.1080/09500690500336957

Topçu, M. S., Mugaloglu, E. Z., & Güven, D. (2014). Socioscientific issues in science education: The case of Turkey. *Educational Sciences: Theory & Practice, 14*(6), 2340. https://doi.org/10.12738/estp.2014.6.2226

Index

R. Khishfe, *The Complexities of Socioscientific Issues in Education*, Springer Texts in Education, https://doi.org/10.1007/978-3-031-88006-3

The manufacturer's authorised representative in the EU is Springer Nature Customer Service Centre GmbH, Europaplatz 3, 69115 Heidelberg, Germany. If you have any concerns regarding our products, please contact ProductSafety@springernature.com

Printed and bound by CPI Group (UK) Ltd, Croydon, CR0 4YY
01/10/2026
02225957-0020